Independence
Theory in
Combinatorics

CHAPMAN AND HALL
MATHEMATICS SERIES

Edited by Professor R. Brown
Head of the Department of Pure Mathematics,
University College of North Wales, Bangor
and Dr M. A. H. Dempster,
University Lecturer in Industrial Mathematics
and Fellow of Balliol College, Oxford

Independence Theory in Combinatorics

An Introductory Account with Applications to Graphs and Transversals

VICTOR BRYANT

and HAZEL PERFECT

Department of Pure Mathematics,
University of Sheffield

1980

LONDON AND NEW YORK

CHAPMAN AND HALL

150TH ANNIVERSARY

First published 1980 by
Chapman and Hall Ltd
11 New Fetter Lane, London EC4P 4EE
Published in the USA by
Chapman and Hall
in association with Methuen, Inc.
733 Third Avenue, New York, NY 10017
© 1980 V. Bryant and H. Perfect

Printed in Great Britain at the
University Press, Cambridge

ISBN 0 412 16220 2
ISBN 0 412 22430 5 (paperback)

British Library Cataloguing in Publication Data

Bryant, Victor
   Independence theory in combinatorics.
   – (Chapman and Hall mathematics series).
   1. Matroids
   I. Title   II. Perfect, Hazel
   511'.6   QA166.6   79-42818

ISBN 0-412-16220-2

Contents

Preface

Combinatorics may very loosely be described as that branch of mathematics which is concerned with the problems of arranging objects in accordance with various imposed constraints. It covers a wide range of ideas and because of its fundamental nature it has applications throughout mathematics. Among the well-established areas of combinatorics may now be included the studies of graphs and networks, block designs, games, transversals, and enumeration problems concerning permutations and combinations, from which the subject earned its title, as well as the theory of independence spaces (or matroids). Along this broad front, various central themes link together the very diverse ideas. The theme which we introduce in this book is that of the abstract concept of independence. Here the reason for the abstraction is to unify; and, as we shall see, this unification pays off handsomely with applications and illuminating sidelights in a wide variety of combinatorial situations.

The study of combinatorics in general, and independence theory in particular, accounts for a considerable amount of space in the mathematical journals. For the most part, however, the books on abstract independence so far written are at an advanced level, whereas the purpose of our short book is to provide an elementary introduction to the subject. It is intended primarily as a text book for undergraduates or as a basis of a postgraduate course and, within its limited compass, we have tried to present the basic notions and to describe just a few of the varied applications of this very attractive branch of mathematics.

We have deliberately chosen to use the term 'independence space', rather than 'matroid' or 'pre-geometry', because we feel that it is the most descriptive label and, therefore, that it has the most universal appeal.

Chapter 1 is concerned with preliminaries and, in particular, it includes two standard forms of independence in a vector space which motivate our abstraction in Chapter 2, where independence spaces are defined. Here various properties of these spaces are described as well as some of the ways in which they may be generated, for example by 'submodular' functions. In Chapter 3 we introduce some classic realizations of independence spaces in graph theory, and we see these two branches of combinatorics each enhancing the other. We meet yet more surprising and beautiful occurrences of independence in Chapter 4, which is devoted to transversal theory. One of our motivations for independence spaces was by means of linear independence in a vector space and, in Chapter 5, we briefly review the problem of the extent to which abstract independence provides a true generalization of linear independence.

As we have already emphasized, this is an elementary and introductory book on independence spaces, and the ninety or so exercises at the ends of the chapters form an integral part of it. These vary in complexity quite considerably, from the routine ones to those which are technically difficult (and are starred). Notes on the exercises and full solutions of most of them are to be found at the end of the book. The pre-requisites for reading this text are fairly minimal: we assume a certain familiarity with the language and algebra of sets and with the usual basic concepts of linear algebra, together with that degree of 'mathematical maturity' normally to be expected of, at any rate, second- and third-year undergraduates in our universities and polytechnics.

The interested reader may wish to use some of the more advanced books cited in the booklist on page 139, for example to trace the origins of the subject in further detail; for, except for the one or two most famous theorems, we have not attributed any results to their originators. Inevitably, a short account such as this can only outline the subject, but we hope that we have provided a foundation which is both interesting in its own right and at the same time a suitable starting point for the more ambitious reader.

Acknowledgements

We wish to thank a number of people who have helped us while we have been writing this book. Professor Leon Mirsky has encouraged us throughout and, in addition, our own text owes much to his book *Transversal Theory*. Indeed, this latter book and also *Matroid Theory* by Dominic Welsh have both influenced our choice of presentation of certain topics.

During the session 1978–79 we gave an MSc course based on this book, and we believe that the text has benefitted from the comments made by the students who attended: we thank them for their help and mention, in particular, Eleanor Smith and Norman Fenton.

We owe especial thanks to Christine Bryant who prepared a meticulous hand-written copy of the final text for the printers. Finally, we are indebted to Messrs. Chapman and Hall: to their readers for their constructive advice, to their editors for their kind and helpful dealings with us throughout, and to the printers for their excellent work.

University of Sheffield   Victor Bryant
February 1980   Hazel Perfect

List of symbols
and abbreviations

General

$A \backslash B$	difference of A and B
$A \triangle B$	symmetric difference of A and B
$\lvert A \rvert$	cardinality of A
$\mathscr{P}(E)$	power set of E
$\{x_1, \ldots, x_n\}_{\neq}$	set of distinct elements $x_1, \ldots, x_n$
$(x_i : i \in I)$	family indexed by I
$\mathfrak{A}, \mathfrak{B}$ etc.	families of sets
$\mathfrak{A}' \subseteq \mathfrak{A}$	$\mathfrak{A}'$ a sub-family of $\mathfrak{A}$
$GF(2)$	field $\{0, 1\}$
$x \cdot y$	product of vectors x and y
$w(x)$	weight of x

Independence spaces

$\mathscr{E}$	independence structure
$(E, \mathscr{E})$	independence space
$I(1)$	hereditary property
$I(2)$	replacement property
$I(2)'$	alternative form of replacement
$B(1), B(2)$	basis properties
$C(1), C(2), C(2)'$	circuit properties
$D(1), D(2), D(3)$	properties of dependence relation
$R(1), R(2), R(3)$	properties of rank
$R(0)$	additional submodular property
$\mathscr{E} \vert E'$	restriction of $\mathscr{E}$ to E'

$\mathscr{E}_{\otimes E \backslash E'}$	contraction of $\mathscr{E}$ away from E'
$\mathscr{E}(E')$	generalization of contracted structure
$\mathscr{E}^*$	dual of $\mathscr{E}$
$\mathscr{E}_1 \oplus \mathscr{E}_2$	direct sum of $\mathscr{E}_1$ and $\mathscr{E}_2$
$\mathscr{E}_1 + \mathscr{E}_2$	sum of $\mathscr{E}_1$ and $\mathscr{E}_2$
$\mathscr{E}(G)$	cycle structure of G
$\mathscr{E}^*(G)$	cutset structure of G
$\mathscr{E}(\mathfrak{A})$	transversal structure of $\mathfrak{A}$
ρ	rank function
$x \mid A$	x depends on A
$\dim X$	dimension of X in a vector space
$\mathscr{C}(\mathscr{E})$	characteristic set of $\mathscr{E}$

Graphs

$G = (V, E)$	graph
$\boldsymbol{G} = (V, \boldsymbol{E})$	directed graph
(E, Δ, E')	bipartite graph
Δ	set of edges of (E, Δ, E')
$\Delta(A)$	subset of vertices of (E, Δ, E')
K_n	complete graph
$K_{r,s}$	complete bipartite graph
G^*	geometric dual of G
$c(A)$	number of components of (V, A)
S_v	star at v
$\mathfrak{S}_X$	family of stars indexed by X

CHAPTER ONE

Preliminaries

1.1 General introductory and historical remarks

Plato is reputed to have said 'God ever geometrizes'. Since geometrical terminology is widely used throughout mathematics, and 'spaces' of all kinds are studied nowadays, mathematicians have evidently followed the divine example. We find it natural to think in visual terms even in branches of mathematics which are not themselves basically geometrical. For example, in elementary mathematics we draw graphs of functions and Venn diagrams of sets; on a higher level, linear equations and linear algebra are clarified when studied within the framework of general 'vector spaces', and calculus and analysis have been transformed by the study of 'metric spaces' and 'topological spaces'. In the last forty years or so 'independence spaces' (called by many authors 'matroids' or 'pre-geometries') have found a place in the mathematical literature, and the insight which they have increasingly brought, notably to parts of algebra, to graph theory and more general combinatorics, make us confident of their continuing importance in the future.

We now give a very brief outline of the historical development of the subject of independence theory: of course, the terms used and the results quoted will mean much more to the reader as he works through the text. In the 1930s the idea of 'abstract independence' emerged principally from two different sources. It is interesting to note some crucial differences between the first two editions of

Moderne Algebra by van der Waerden in which he treats two types of dependence, linear and algebraic. In the first edition the two subjects are dealt with quite separately whereas, in the later account, van der Waerden lists just three basic properties of a relation of dependence and proceeds to show that they imply all that is needed for a study of both types of dependence. From quite another direction, the idea of some basic axioms of independence appeared in 1935 in the work of H. Whitney (who introduced the term matroid). His fundamental paper on the subject was motivated by his earlier work in graph theory. The names of Garrett Birkhoff and Saunders MacLane are also associated with the early developments of the subject, mainly from the point of view of lattices. New impetus came in the 1940s when Richard Rado generalized Philip Hall's classic theorem on 'distinct representatives' to a parallel theorem on 'independent representatives', thus forging the links between the theory of independence and 'transversal theory'. More recently, the work of W. T. Tutte in the development of independence spaces has been of profound significance, and it is to him that we owe, for example, the characterization theorems for graphic spaces. The subject continues to flourish today as new combinatorial realizations of independence spaces come to light and, with them, an ever-increasing range of applications.

1.2 Sets, families and graphs

We take for granted that the reader is familiar with the essentials of the elementary algebra of sets, and we mention here only one or two points of notation which are liable to vary from text to text. We shall occasionally refer explicitly to the set $\{0, 1, 2, \ldots\}$ of non-negative integers, and to vector spaces over an arbitrary field, but otherwise ALL SETS WHICH OCCUR WILL BE FINITE. This assumption is made for reasons of simplicity and brevity rather than of necessity. If A, B are sets and A is a subset of B we shall write $A \subseteq B$; if in addition $A \neq B$, then we say that A is a *proper* subset of B and indicate this by writing $A \subset B$. Our notation for the *difference* of A and B is $A \backslash B$: this is the set of those elements in A but not in B. By the *symmetric difference* of A and B we mean the set $(A \backslash B) \cup (B \backslash A)$ (or $(A \cup B) \backslash (A \cap B)$) and we write this as $A \triangle B$. The operation $\triangle$ is associative, and the set $A_1 \triangle \ldots \triangle A_n$ consists of those elements which belong to precisely an odd number of the A_is. When we consider $A = \{x_1, \ldots, x_n\}$, and we wish to emphasize that $x_1, \ldots, x_n$ are all different, we shall write $A = \{x_1, \ldots, x_n\}_{\neq}$. We shall denote the number of elements (or *cardinality*) of the set A by $|A|$. The letter E is often reserved for an underlying set whose subsets we consider. The collection of all subsets of E forms the *power set* of E, and will be denoted by $\mathscr{P}(E)$.

In our subject (and particularly in Chapter 4) it is of especial importance to distinguish the notion of a 'family' from that of a 'set'. It is perhaps helpful to look first at an example: the sets $\{1, 2\}$, $\{2, 1\}$, $\{1, 1, 2\}$ are all equal (i.e. are one and the same set written in different ways) whereas $(1, 2)$, $(2, 1)$, $(1, 1, 2)$ are three different families. We emphasize the difference notationally by the use of different kinds of brackets. The point which this example illustrates is that two sets are the same if they contain just the same elements; whereas in the more refined notion of a family the ordering of the elements and the number of times each occurs are both significant. More formally, a typical *family* $(x_i : i \in I)$ of elements x_i of a set E is a mapping $x : I \to E$ in which x_i denotes the image of i under x. If $I' \subseteq I$, then the restriction of x to I', $(x_i : i \in I')$ is called a *subfamily* of $(x_i : i \in I)$. Since, for our purposes, all sets are finite there is no essential loss of generality in taking the index set I to be $\{1, \ldots, n\}$ and writing the n-tuple $(x_1, \ldots, x_n)$ instead of $(x_i : i \in I)$; finite families are indeed just finite sequences. We shall feel free to use either notation. Particularly in Chapter 4, we shall also encounter families $(A_1, \ldots, A_n)$ of subsets of a given set E; these, then, are mappings $A : \{1, \ldots, n\} \to \mathscr{P}(E)$, where A_i denotes the image of i under A.

We now turn briefly to graphs, where our interest is two-fold: graphs are the main object of study throughout Chapter 3; but, before then, graphs of a particularly simple kind, namely 'bipartite' graphs, play an important subsidiary role in our general development of the theory of independence spaces. A graph $G = (V, E)$ consists essentially of a non-empty set V of *vertices* and a set E of *edges* 'joining' certain pairs of vertices; thus, with each $e \in E$, there are associated elements $v, w \in V$ called the *endpoints* of e, and e is said to *join* v and w. We allow multiple edges (in the sense that several edges can join the same pair of vertices) and loops (which are edges whose two endpoints coincide). If e joins v and w, then unless the graph has multiple edges we can write $e = vw$ (or wv) unambiguously. The *degree* of a vertex v in G is the number of ends of edges of G at v. (This is the same as the number of edges of G which have v as an endpoint, where a loop through v counts as two edges.)

Examples

It is easy to draw representations of graphs in an obvious fashion, the vertices being represented by points and the edges by lines joining the relevant pairs.

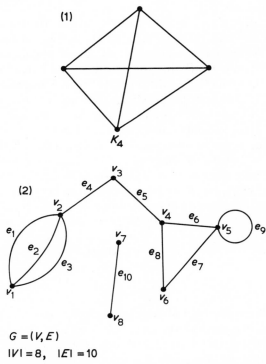

The graph (1) consists of four vertices, no loops, and every pair of distinct vertices joined by one edge; it is called K_4, the *complete graph* on four vertices. Each vertex has degree 3. The complete graph K_n on n vertices is defined in a similar way. The graph (2) has multiple edges and just one loop. We note that the degree of the vertex v_5 is equal to 4.

Terminology in graph theory is still far from standard. In what follows we shall understand by a *path* in a graph a sequence of the form

$$v_1, e_1, v_2, e_2, \ldots, e_n, v_{n+1},$$

where the $v_1, \ldots, v_{n+1}$ are distinct vertices and each e_i joins v_i and v_{i+1}. In particular, for each $v \in V$, the sequence (v) is a 'degenerate' path. (The case $n = 0$.) However, having made the point about distinct vertices, for most purposes we can regard the vertices in the sequence as redundant and specify a non-degenerate path simply by the set $\{e_1, \ldots, e_n\}$ of its edges; it has *endpoints* (and it *joins*) v_1 and v_{n+1}. Thus, in Example 2 above, $\{e_1, e_4, e_5\}$ is a path joining v_1 and v_4. Similarly, a *cycle* (occasionally called a closed path) is a sequence

$$v_1, e_1, v_2, e_2, \ldots, e_n, v_1,$$

where now $n \geqslant 1$, the $v_1, \ldots, v_n$ are distinct vertices, each $e_i (1 \leqslant i \leqslant n - 1)$ joins v_i and v_{i+1}, and e_n joins v_n and v_1. Again, it can be specified more simply by the (non-empty) set $\{e_1, \ldots, e_n\}$ of its edges. In Example 2 above, $\{e_6, e_7, e_8\}$ and $\{e_9\}$ are both cycles. A graph is *connected* if every two vertices are the endpoints of a path in the graph. Given a graph $G = (V, E)$, it is possible to partition V and E into $V = V_1 \cup \ldots \cup V_c$ and $E = E_1 \cup \ldots \cup E_c$ such that the $V_1, \ldots, V_c$ are non-empty and each of $(V_1, E_1), \ldots, (V_c, E_c)$ is a connected graph; these are called the (connected) *components* of G. Note that a component may consist of a single vertex. In Example 1, K_4 has only one component and so is connected, whereas, in Example 2, G has two components. A graph without cycles is called a *forest*; if, in addition, it is connected it is called a *tree* (if the reader draws a few such graphs he may see why). For simplicity, the edge-sets of forests and trees are themselves also called forests and trees. A set of edges in a graph G, no two of which have a common endpoint, is called a *matching* in G. In Example 2, the set $\{e_4, e_6, e_{10}\}$ is a matching. Of especial importance in Chapter 2 are bipartite graphs. A graph G without multiple edges is called *bipartite* if its vertex set can be partitioned into two non-empty sets, say X and Y, such that every edge of G joins a vertex of X and a vertex of Y.

Examples

The graphs 3 and 4 overleaf are both bipartite. The graph 4 is the *complete bipartite graph* $K_{3,4}$; more generally, the graph $K_{r,s}$ can be defined in the obvious fashion.

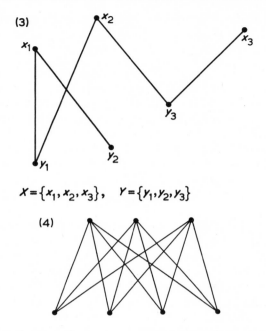

$$X=\{x_1, x_2, x_3\}, \quad Y=\{y_1, y_2, y_3\}$$

Other graph-theoretic notions will be introduced as and when necessary.

1.3 Vector spaces; linear and affine independence

One of the simplest motivations for the general notion of 'independence' is provided by the study of vector spaces. We shall assume that the reader has encountered the rudiments of the theory of finite-dimensional vector spaces over an arbitrary field. Here we remind him especially of two important 'replacement theorems'. Recall first that a set $A = \{u_1, \ldots, u_m\}_{\neq}$ of vectors is *linearly independent* over a field F if, whenever $\alpha_1 u_1 + \ldots + \alpha_m u_m = 0$ (the zero vector) for some $\alpha_i \in F$, it follows that $\alpha_1 = \ldots = \alpha_m = 0$; otherwise A is *linearly dependent*.

Theorem 1.1. (The Replacement Theorem for linear independence) *If A, B are linearly independent subsets of a vector space with $|B| = |A| + 1$, then there exists $v \in B \backslash A$ such that $A \cup \{v\}$ is linearly independent.*

Proof. Let the underlying field be F. Assume that the result fails for $A = \{u_1, \ldots, u_m\}_{\neq}$ and $B = \{v_1, \ldots, v_{m+1}\}_{\neq}$. Then, for each i with $1 \leqslant i \leqslant m+1$, either $v_i \in \{u_1, \ldots, u_m\}$ or $\{u_1, \ldots, u_m, v_i\}_{\neq}$ is linearly dependent. It readily follows that there exist elements $\alpha_{ij} \in F$ such that

$$v_1 = \alpha_{11} u_1 + \ldots + \alpha_{m1} u_m,$$

$$\cdots$$

$$v_{m+1} = \alpha_{1, m+1} u_1 + \ldots + \alpha_{m, m+1} u_m.$$

Further, it is well known that $m+1$ elements $\beta_1, \ldots, \beta_{m+1} \in F$, not all zero, can always be found to satisfy the m equations

$$\alpha_{11} \beta_1 + \ldots + \alpha_{1, m+1} \beta_{m+1} = 0,$$

$$\cdots$$

$$\alpha_{m1} \beta_1 + \ldots + \alpha_{m, m+1} \beta_{m+1} = 0;$$

and therefore

$$\sum_{i=1}^{m+1} \beta_i v_i = \sum_{i=1}^{m+1} \beta_i \left(\sum_{j=1}^{m} \alpha_{ji} u_j \right)$$

$$= \sum_{j=1}^{m} \left(\sum_{i=1}^{m+1} \alpha_{ji} \beta_i \right) u_j = 0.$$

This, however, contradicts the linear independence of $\{v_1, \ldots, v_{m+1}\}_{\neq}$; and we conclude that the Replacement Theorem holds. $\qquad\square$

Now let V be a vector space (over a field F) which possesses a finite maximal linearly independent set, i.e. a finite linearly independent set not properly contained in any other (V is then called *finite dimensional*), and let X be a subset of V. A consequence of Theorem 1.1 is that all maximal linearly independent subsets, or *bases*, of X have the same finite number of elements. For, given any basis B of X and any linearly independent subset A of X of fewer elements, it is clear from the theorem that A cannot be maximal. This common cardinality of the bases of X is called the *dimension* or *linear rank*, of X and is denoted by $\dim X$. In particular, these results apply to $X = V$ itself. Vector spaces of dimensions 1 and 2 are generally called lines and planes, respectively. The *linear subspaces* of V are precisely those non-empty subsets X of V with the property that, whenever $x, y \in X$ and $\alpha, \beta \in F$, then $\alpha x + \beta y \in X$. This turns out to be equivalent to saying that, whenever $v \in V \backslash X$, then $\dim (X \cup \{v\}) > \dim X$. Clearly, every linear subspace of V contains the zero vector of V and is itself a vector space over F.

Closely related to the notion of linear independence in a vector space is that of 'affine independence'. In this latter notion, however, the origin or zero vector ceases to play a special role. Let, then, V be again a vector space over a field F. A set $A = \{u_1, \ldots, u_m\}_{\neq}$ of members of V is said to be *affinely independent* over F if, whenever $\alpha_1 u_1 + \ldots + \alpha_m u_m = 0$ for some $\alpha_i \in F$ with $\alpha_1 + \ldots + \alpha_m = 0$, then $\alpha_1 = \ldots = \alpha_m = 0$. It is easy to check that, if $A = \{u_1, \ldots, u_m\}_{\neq}$ is affinely independent, then so is any 'translate' $\{u_1 - u, \ldots, u_m - u\}$ of A; and that A is affinely independent if and only if $\{u_1 - u_m, \ldots, u_{m-1} - u_m\}$ is linearly independent. In particular, in the ordinary real plane, any three vectors whose endpoints are non-collinear are affinely independent, as are any four vectors in three dimensions whose endpoints are non-coplanar. (More briefly we say that any three non-collinear points, or any four non-coplanar points, are affinely independent.) Because of the very close link between linear and affine independence we do not dwell long on the latter concept; we do, however, deduce the analogue of Theorem 1.1.

Theorem 1.2. (The Replacement Theorem for affine independence). *If A, B are affinely independent subsets of a vector space V with $|B| = |A| + 1$, then there exists $v \in B \setminus A$ such that $A \cup \{v\}$ is affinely independent.*

Proof. We observe first that, if $C = \{w_1, \ldots, w_k\}_{\neq}$ (where $k \geqslant 2$) is affinely independent, then $C \setminus \{w_i\}$ is linearly independent for some i. For, if not, the set C is contained in a $(k-2)$-dimensional linear subspace U of V. This implies, in particular, that also each of the vectors $w_1 - w_k, \ldots, w_{k-1} - w_k$ belongs to U and so $\{w_1 - w_k, \ldots, w_{k-1} - w_k\}_{\neq}$ cannot be linearly independent. But then C cannot be affinely independent.

Now write $A = \{u_1, \ldots, u_m\}_{\neq}$, $B = \{v_1, \ldots, v_{m+1}\}_{\neq}$ and assume that A and B are affinely independent. Then $A' = \{u_1 - u_m, \ldots, u_{m-1} - u_m\}_{\neq}$ is linearly independent, $\{v_1 - u_m, \ldots, v_{m+1} - u_m\}_{\neq}$ is affinely independent and, by the above argument, we may assume that $B' = \{v_1 - u_m, \ldots, v_m - u_m\}_{\neq}$ is linearly independent. Theorem 1.1 applied to A' and B' now shows that $\{u_1 - u_m, \ldots, u_{m-1} - u_m, v_i - u_m\}_{\neq}$ is linearly independent for some i; and hence that $\{u_1, \ldots, u_m, v_i\}_{\neq}$ is affinely independent. $\qquad\square$

Thus, as before, any two maximal affinely independent subsets of X in V have the same cardinality, called the *affine rank* of X. Every singleton subset of V has affine rank 1 (there is nothing special about the zero vector) and every doubleton subset affine rank 2. A subset X of V is an *affine subspace* of V if and only if either $X = \phi$, or $X \neq \phi$ and for $u \in X$ the set $\{x - u : x \in X\}$ is a linear subspace of V; i.e. the non-empty affine subspaces of V are the translates of linear subspaces. Affine independence will provide us with occasional useful examples in the later chapters.

We have now accomplished the main purpose of this section by establishing the Replacement Theorems. However, vector spaces will be important to us, not only because they motivate the abstract concept of independence, but also in connection with the question of the extent to which our abstract concept resembles linear independence. Particular cases of this 'linear representability' are described in Chapters 3 and 4; and Chapter 5 is devoted to the general problem.

Our final results in this section are of a rather special nature and will be needed in Chapter 3; they are included here for completeness. Given a field F, let F^n be the vector space of n-tuples of members of F, i.e. rows (or columns) of n members of F. It is clear that $\dim F^n = n$. For $x = (\alpha_1, \ldots, \alpha_n)$, $y = (\beta_1, \ldots, \beta_n) \in F^n$ we shall write $x \cdot y = \sum_{i=1}^{n} \alpha_i \beta_i (\in F)$ and call it the *product* of x and y.

Theorem 1.3. *Let X be a linear subspace of F^n, and define $Y \subseteq F^n$ by the rule*

$$Y = \{y \in F^n : y \cdot x = 0 \quad \text{for all} \quad y \in Y\}.$$

Then Y is also a linear subspace of F^n and

$$\dim X + \dim Y = n.$$

Proof. Certainly $Y \neq \phi$ since $0 \in Y$. Next, if $y, z \in Y$ and $\alpha, \beta \in F$ then, for each $x \in X$,

$$(\alpha y + \beta z) \cdot x = \alpha(y \cdot x) + \beta(z \cdot x) = 0,$$

and so $\alpha y + \beta z \in Y$; and we conclude that Y is a linear subspace of F^n. Now if X has a basis $\{x_1, \ldots, x_r\}_{\neq}$, then this linearly independent set is contained in a maximal linearly independent subset, or basis, of F^n; $\{x_1, \ldots, x_n\}_{\neq}$ say. Let M be the $n \times n$ (non-singular) matrix with rows $x_1, \ldots, x_n$ (for each x_i can be regarded as a row of n members of F) and let $y_1, \ldots, y_n$ be the columns of the inverse of M. Then $y_1, \ldots, y_n$ can be regarded as members of F^n, and they have the property that

$$x_i \cdot y_j = \begin{cases} 1 & \text{if} \quad i = j \\ 0 & \text{if} \quad i \neq j \end{cases}$$

Also $\{y_1, \ldots, y_n\}_{\neq}$ is linearly independent, and hence a basis of F^n. Now let $y \in F^n$ and write $y = \mu_1 y_1 + \ldots + \mu_n y_n$ ($\mu_i \in F$). Then

$$y \in Y \Leftrightarrow (\mu_1 y_1 + \ldots + \mu_n y_n) \cdot x = 0 \text{ for all } x \in X$$
$$\Leftrightarrow (\mu_1 y_1 + \ldots + \mu_n y_n) \cdot x_i = 0 \quad \text{for} \quad 1 \leqslant i \leqslant r$$
$$\Leftrightarrow \mu_1 = \ldots = \mu_r = 0$$
$$\Leftrightarrow y = \mu_{r+1} y_{r+1} + \ldots + \mu_n y_n.$$

Hence $\{y_{r+1}, \ldots, y_n\}_{\neq}$ is a basis of Y, and $\dim Y = n - r = n - \dim X$. $\square$

Corollary 1.4. *Let X be a linear subspace of F^n, and let*

$$Y = \{y \in F^n : y \cdot x = 0 \quad \text{for all} \quad x \in X\}.$$

Then

$$X = \{x \in F^n : x \cdot y = 0 \quad \text{for all} \quad y \in Y\}.$$

Proof. Let $Z = \{x \in F^n : x \cdot y = 0 \text{ for all } y \in Y\}$. Then, if $x \in X$, it follows that $x \cdot y = 0$ for all $y \in Y$, and so $x \in Z$. Thus $X \subseteq Z$. Now, by Theorem 1.3 applied to X, Y and to Y, Z in turn,

$$\dim X + \dim Y = n$$

and

$$\dim Y + \dim Z = n,$$

and hence $\dim X = \dim Z$. Since, again by Theorem 1.3, X and Z are both linear subspaces, it follows that $X = Z$. $\square$

In the case when F is the real field, what we have called a 'product' is usually called an 'inner product' or 'scalar product', and the spaces

X and Y above are orthogonal complements of each other. For arbitrary F, however, it is not always true that $X \cap Y = \{0\}$ and $X + Y = F^n$ (see Chapter 3). This more general situation, described in simple terms above, is more usually presented in the context of a vector space and its 'dual' space of linear functionals.

Exercises

(V is a finite-dimensional vector space over a field F.)

1.1 Prove that a non-empty subset X of V is a linear subspace of V if and only if $\dim(X \cup \{v\}) > \dim X$ for every $v \in V \backslash X$.

1.2 Prove that any translate of an affinely independent subset of V is itself affinely independent.

1.3 Prove that a subset $\{u_1, \ldots, u_m\}_{\neq}$ of V is affinely independent if and only if $\{u_1 - u_m, \ldots, u_{m-1} - u_m\}_{\neq}$ is linearly independent.

1.4 Let $X \subseteq V$. Prove that X is an affine subspace of V if and only if affine rank $(X \cup \{v\}) > $ affine rank X for every $v \in V \backslash X$.

1.5 Prove that the subset $\{(\alpha_1^1, \ldots, \alpha_n^1), \ldots, (\alpha_1^r, \ldots, \alpha_n^r)\}_{\neq}$ of F^n is affinely independent if and only if the subset $\{(\alpha_1^1, \ldots, \alpha_n^1, 1), \ldots, (\alpha_1^r, \ldots, \alpha_n^r, 1)\}_{\neq}$ of F^{n+1} is linearly independent. Use this fact to give an alternative proof of the Replacement Theorem for affine independence in V.

CHAPTER TWO

Independence
spaces

2.1 Axioms and some basic theorems

From among many possible starting points, we choose to define independence spaces by means of axioms ($I_{(1)}$, $I_{(2)}$ below) which immediately reflect the main properties of linearly independent sets in a vector space.

Let E be a given set and $\mathscr{E}$ a non-empty collection of subsets of E with the following properties

I(1) $\mathscr{E}$ is 'hereditary'; i.e. if $A \in \mathscr{E}$ and $B \subseteq A$, then $B \in \mathscr{E}$;

I(2) $\mathscr{E}$ has the property of 'replacement'; i.e. if $A, B \in \mathscr{E}$ with $|B| = |A| + 1$, then there exists $x \in B \backslash A$ such that $A \cup \{x\} \in \mathscr{E}$.

Then $\mathscr{E}$ is called an *independence structure* on E and its members *independent sets*. (We observe that $\phi \in \mathscr{E}$.) The pair $(E, \mathscr{E})$ is called an *independence space*. Subsets of E which are not independent are *dependent*.

A set of vectors together with the collection of its linearly independent subsets provides the first obvious example of an independence space. Half of this book will be devoted to two other classes of examples, much less obvious: graphic spaces in Chapter 3 and transversal spaces in Chapter 4. In passing, we mention the trivial example of the independence space $(E, \mathscr{E})$, where $\mathscr{E} = \{\phi\}$, and, at the other end of the scale, the independence space $(E, \mathscr{E})$, where $\mathscr{E} = \mathscr{P}(E)$, the collection of all subsets of E. In this latter situation, we speak of $\mathscr{E}$

as the *universal structure* on E. Note that the repeated application of axiom I(2) easily gives the following extension:

I(2)′ If $A, B \in \mathscr{E}$ with $|A| \leq |B|$, then there exists $A' \in \mathscr{E}$ with $A \subseteq A' \subseteq A \cup B$ and $|A'| = |B|$.

Since, evidently, I(2)′ implies I(2), it follows that I(2) and I(2)′ are equivalent and, in future, we shall refer to them both unambiguously as I(2).

A maximal independent set in an independence space $(E, \mathscr{E})$, that is to say one which is not properly contained in an independent set, will be called a *basis* of $(E, \mathscr{E})$, or of $\mathscr{E}$.

Theorem 2.1. *Any two bases of an independence space contain the same number of elements.*

Proof. Suppose B and B' are bases of the independence space $(E, \mathscr{E})$ with $|B| < |B'|$. Then, by I(2), there exists $A' \in \mathscr{E}$ with $B \subseteq A' \subseteq A \cup B'$ and $|A'| = |B'|$. But then $B \subset A'$; which contradicts the maximality of B. $\qquad\qquad\square$

Theorem 2.2. *The bases of an independence space have the following properties*

B(1) *No proper subset of a basis is a basis*;

B(2) *if B, B' are bases and $x \in B$, then $(B \backslash \{x\}) \cup \{y\}$ is also a basis for some $y \in B'$.*

Proof. The truth of B(1) follows at once from the maximality of a basis. Next, if B, B' are bases of $(E, \mathscr{E})$ and $x \in B$, then $B \backslash \{x\} \in \mathscr{E}$ and $|B'| = |B \backslash \{x\}| + 1$, by Theorem 2.1. So, by I(2), there exists $y \in B' \backslash (B \backslash \{x\})$ such that $(B \backslash \{x\}) \cup \{y\} \in \mathscr{E}$. Also $|(B \backslash \{x\}) \cup \{y\}| = |B|$ and so, again by Theorem 2.1, $(B \backslash \{x\}) \cup \{y\}$ is a maximal independent set as required. $\qquad\qquad\square$

We devote the following section to a discussion of some induced structures. It is useful, however, to introduce a particularly simple one here. Given any subset A of E in an independence space $(E, \mathscr{E})$ we can consider those members of $\mathscr{E}$ which are contained in A. In fact, it is easy to check that these sets themselves form an independence structure on A; we call it the *restriction* of $\mathscr{E}$ to A and denote it by $\mathscr{E}|A$.

An immediate consequence is that the properties of bases of an independence space just established apply equally well to the maximal independent subsets (in $\mathscr{E}$) of a set A (in E). Hence we are led naturally to the notion of rank: the *rank* of A in $(E, \mathscr{E})$ is the common cardinality of all maximal independent subsets of A. We shall denote it by $\rho(A)$. We call ρ the *rank function* of $\mathscr{E}$; $\rho(E)$ is itself also called the rank of the space $(E, \mathscr{E})$ or of $\mathscr{E}$.

Theorem 2.3. *The rank function ρ of an independence space $(E, \mathscr{E})$ has the following properties for all subsets of the space*

R(1) $0 \leqslant \rho(A) \leqslant |A|$;

R(2) *if* $B \subseteq A$, *then* $\rho(B) \leqslant \rho(A)$;

R(3) $\rho(A) + \rho(B) \geqslant \rho(A \cup B) + \rho(A \cap B)$.

Proof. The truth of R(1) is immediate since $\rho(A)$ is the cardinality of some subset of A. Similarly, R(2) is very easy since, if $B \subseteq A$, then a maximal independent subset of B is contained in a maximal independent subset of A. To establish R(3), let X be a maximal independent subset of $A \cap B$. This may be extended to a maximal independent subset $X \cup Y$ of A and then further to a maximal independent subset $X \cup Y \cup Z$ of $A \cup B$, where $Y \subseteq A \backslash B$ and $Z \subseteq B \backslash A$. Evidently, then

$$\rho(A \cap B) = |X|, \quad \rho(A) = |X| + |Y|,$$
$$\rho(A \cup B) = |X| + |Y| + |Z|$$

and, since $X \cup Z \in \mathscr{E}$ and $X \cup Z \subseteq B$, we have

$$\rho(B) \geqslant |X \cup Z| = |X| + |Z| = \rho(A \cap B) + (\rho(A \cup B) - \rho(A));$$

and the desired result follows at once. $\qquad \square$

For our later applications, property R(3), known as the 'submodular inequality', will be of especial importance. We note that it cannot in general be replaced by an equality; for example, if $E = \{1, 2, 3\}$ and $\mathscr{E} = \{\phi, \{1\}, \{2\}, \{3\}, \{1, 2\}, \{2, 3\}\}$, then $(E, \mathscr{E})$ is an independence space, but if $A = \{1\}$ and $B = \{3\}$, then

$$2 = \rho(A) + \rho(B) > \rho(A \cup B) + \rho(A \cap B) = 1.$$

The next important concept which we introduce is that of a *circuit*, which is a minimal dependent set in an independence space. In other words, $C \subseteq E$ is a circuit of $(E, \mathscr{E})$ if and only if $C \notin \mathscr{E}$ but $C\backslash\{x\} \in \mathscr{E}$ for each $x \in C$.

Theorem 2.4. *The circuits of an independence space have the following properties*

    C(1) *No proper subset of a circuit is a circuit*;

    C(2) *if C, C' are circuits with $x \in C \cap C'$, $y \in C\backslash C'$, then there exists a circuit C^* such that $y \in C^* \subseteq (C \cup C')\backslash\{x\}$.*

Proof. The truth of C(1) follows at once from the minimality of a circuit. In order to prove C(2), let C, C' be circuits of $(E, \mathscr{E})$. Then $C'\backslash\{x\} \in \mathscr{E}$ and $C'\backslash\{x\} \subseteq (C \cup C')\backslash\{y\}$, and so there exists a maximal independent subset B of $(C \cup C')\backslash\{y\}$ with $C'\backslash\{x\} \subseteq B$. Since $C' \notin \mathscr{E}$, it follows that $x \notin B$ and hence that B is also a maximal independent subset of $(C \cup C')\backslash\{x, y\}$. Similarly, if B' is a maximal independent subset of $C \cup C'$ with $C\backslash\{y\} \subseteq B'$, then $y \notin B'$ and B' is a maximal independent subset of $(C \cup C')\backslash\{y\}$. Hence, if ρ is the rank function of $\mathscr{E}$,

$$\rho(C \cup C') = \rho((C \cup C')\backslash\{y\}) = \rho((C \cup C')\backslash\{x, y\})$$

and, since $(C \cup C')\backslash\{x, y\} \subseteq (C \cup C')\backslash\{x\} \subseteq C \cup C'$, therefore

$$\rho((C \cup C')\backslash\{x\}) = \rho((C \cup C')\backslash\{x, y\}).$$

So $B \cup \{y\}$ is a dependent subset of $(C \cup C')\backslash\{x\}$ which contains y, and the removal of y makes it independent. Therefore $B \cup \{y\}$ contains a circuit C^* such that $y \in C^* \subseteq (C \cup C')\backslash\{x\}$.   □

We began this section by remarking that, in stating the axioms for independent sets, we had chosen one of many possible starting points. Axioms for bases, circuits, rank, etc., provide alternative approaches. We do not wish to go into the matter of the very considerable number of different axiom systems in any great detail, but in some of our later examples it is easier, for instance, to start with bases or circuits than with independent sets. So we now establish that the properties B(1) and B(2) of Theorem 2.2 characterize the bases of an independence space, and that C(1) and C(2) of Theorem 2.4 characterize the circuits. We shall also see later, in connection with our study of submodular functions in Section 2.3, that the properties R(1), R(2) and R(3) are indeed characteristic properties of the rank function in an independence space.

Theorem 2.5. *Let E be a given set, and let $\mathscr{B}$ be a non-empty collection of subsets of E which satisfy the conditions*

  B(1) *No proper subset of a member of $\mathscr{B}$ is itself a member of $\mathscr{B}$;*
  B(2) *if B, $B' \in \mathscr{B}$ and $x \in B$, then $(B \setminus \{x\}) \cup \{y\} \in \mathscr{B}$ for some $y \in B'$.*

Let $\mathscr{E}$ be the collection of subsets of E defined by the condition that $X \in \mathscr{E}$ if and only if X is a subset of some member of $\mathscr{B}$. Then $(E, \mathscr{E})$ is an independence space, and $\mathscr{B}$ is precisely its collection of bases.

Proof. It is at once clear that $\mathscr{E} \neq \phi$ and that $\mathscr{E}$ satisfies I(1). We next observe that, if $A \in \mathscr{E}$ with A contained in a member B' of $\mathscr{B}$ of cardinality n say, and if $B \in \mathscr{B}$, then there exists $B^* \in \mathscr{B}$ with $A \subseteq B^* \subseteq A \cup B$ and $|B^*| \leqslant n$. For we may choose $B^* \in \mathscr{B}$ with $A \subseteq B^*$, $|B^*| \leqslant n$ and $|B^* \setminus B|$ minimal; and an easy application of B(2) to B^* and B shows that, unless $B^* \setminus (A \cup B) = \phi$, the minimality assumption is contradicted. Therefore $B^* \subseteq A \cup B$. From this fundamental fact, we are able to deduce two consequences: first, that all members of $\mathscr{B}$ have the same cardinality and, second, that $\mathscr{E}$ satisfies I(2).

If $B, B' \in \mathscr{B}$ and $|B'| < |B|$, then $B' \cap B \in \mathscr{E}$ and there exists $B^* \in \mathscr{B}$ with $B' \cap B \subseteq B^* \subseteq (B' \cap B) \cup B = B$ and $|B^*| \leqslant |B'| < |B|$. But then $B^* \subset B$; which contradicts B(1).

Next, given $A, A' \in \mathscr{E}$ with $|A'| = |A| + 1$, there exists $B \in \mathscr{B}$ with $A' \subseteq B$, and hence also $B^* \in \mathscr{B}$ with $A \subseteq B^* \subseteq A \cup B$. Therefore $B^* \setminus A$ and A' are both subsets of B and, since

$$|B| = |B^*| < |B^*| - |A| + |A'| = |B^* \setminus A| + |A'|,$$

there exists $x \in (B^* \setminus A) \cap A'$. But then, clearly, $x \in A' \setminus A$ and $A \cup \{x\} \in \mathscr{E}$ since $A \cup \{x\} \subseteq B^*$. Thus $\mathscr{E}$ satisfies I(2).

Finally, it is readily seen that the collection of bases of $\mathscr{E}$ is precisely $\mathscr{B}$.   □

Theorem 2.6. *Let E be a given set, and let $\mathscr{C}$ be a collection of non-empty subsets of E which satisfy the conditions*

  C(1) *No proper subset of a member of $\mathscr{C}$ is itself a member of $\mathscr{C}$;*
  C(2)′ *if $C, C' \in \mathscr{C}, C \neq C'$, and $x \in C \cap C'$, then there exists $C^* \in \mathscr{C}$ such that $C^* \subseteq (C \cup C') \setminus \{x\}$.*

Let $\mathscr{E}$ be the collection of subsets of E defined by the condition that $X \in \mathscr{E}$ if and only if X contains no member of $\mathscr{C}$ as a subset. Then $(E, \mathscr{E})$ is an independence space, and $\mathscr{C}$ is precisely its collection of circuits.

Proof. Since the $\mathscr{E}$ defined above clearly contains ϕ as a member, it follows that $\mathscr{E} \neq \phi$. Also I(1) is immediately satisfied for the collection $\mathscr{E}$; for if a set A contains no member of $\mathscr{C}$, and $B \subseteq A$, then B contains no member of $\mathscr{C}$. So it remains to establish I(2) in order to show that $\mathscr{E}$ is an independence structure on E. To this end, let $A, B \in \mathscr{E}$ with $|A| = m$, $|B| = m + 1$. So there exists a subset of $A \cup B$ of cardinality $m + 1$ which is a member of $\mathscr{E}$; let B' be such a set chosen with $|A \backslash B'|$ minimal. We aim to show that $A \backslash B' = \phi$ (so that $A \subset B'$). Assume the contrary; that there exists $x \in A \backslash B'$. For each $y \in B' \backslash A$, the set $B_y = (\{x\} \cup B') \backslash \{y\}$ has cardinality $m + 1$, it is a subset of $A \cup B$, and $|A \backslash B_y| < |A \backslash B'|$. Hence $B_y \notin \mathscr{E}$ and so it contains a member C_y of $\mathscr{C}$ as a subset. Evidently $x \in C_y$ and $y \notin C_y$. If each $y \in B' \backslash A$ gives rise to the same member of $\mathscr{C}$, C say, then for each $y \in B' \backslash A$ it follows that $y \notin C$. But then $C \subseteq A$, which is impossible since $A \in \mathscr{E}$. Hence C_{y_1} and C_{y_2}, say, are different members of $\mathscr{C}$ constructed as above. Thus $x \in C_{y_1} \cap C_{y_2}$ and, by the given property $C(2)'$, there exists a member C^* of $\mathscr{C}$ in $(C_{y_1} \cup C_{y_2}) \backslash \{x\}$ $(\subseteq B' \in \mathscr{E})$. This contradiction shows that $A \subset B'$, and that I(2) is satisfied.

Let us temporarily denote by $\mathscr{C}'$ the collection of circuits of the independence space $(E, \mathscr{E})$. Now

$$C \in \mathscr{C}' \Leftrightarrow C \notin \mathscr{E} \text{ and } C \backslash \{x\} \in \mathscr{E} \text{ for each } x \in C$$

$$\Leftrightarrow \text{there exists } C' \subseteq C \text{ with } C' \in \mathscr{C}, \text{ and}$$
$$\text{for each } x \in C \text{ there is no } C_x \in \mathscr{C}$$
$$\text{with } C_x \subseteq C \backslash \{x\}$$

$$\Leftrightarrow C \in \mathscr{C} \text{ (by } C(1)).$$

Therefore $\mathscr{C}' = \mathscr{C}$, and the collection of circuits of $(E, \mathscr{E})$ is precisely $\mathscr{C}$.

$\square$

It is clear from Theorems 2.4 and 2.6 that the circuits of an independence space satisfy C(2) and C(2)' and that either of these properties together with C(1) characterizes them. We shall therefore refer to either of the pairs C(1), C(2) or C(1), C(2)' as the circuit axioms.

Finally in this section we mention briefly those subsets of E which resemble the subspaces (linear or affine) of a vector space; these are the 'flats' of an independence space $(E, \mathscr{E})$. Recall that the adjunction of an external vector to a subspace increases its dimension or rank; so if the independence space $(E, \mathscr{E})$ has rank function ρ, then a subset A of E is called a *flat* of $(E, \mathscr{E})$ if and only if $\rho(A \cup \{x\}) > \rho(A)$ for every $x \in E\backslash A$.

Theorem 2.7. *Any intersection of flats is itself a flat.*

Proof. Consider $(E, \mathscr{E})$ with rank function ρ. Now let $A_1, \ldots, A_k$ be flats, write $A = A_1 \cap \ldots \cap A_k$, let $x \in E\backslash A$ and suppose that, say, $x \in E\backslash A_i$. Then it remains to show that $\rho(A \cup \{x\}) > \rho(A)$. Let B be a maximal independent subset of $A\ (\subseteq A_i)$ and, by I(2), extend B to $B \cup B'$, a maximal independent subset of A_i with $B' \subseteq A_i\backslash A$. Since A_i is a flat and $x \in E\backslash A_i$, it follows that $\rho(A_i \cup \{x\}) > \rho(A_i)$ and that $B \cup B'$ is not a maximal independent subset of $A_i \cup \{x\}$. Thus $B \cup B' \cup \{x\} \in \mathscr{E}$, and so $B \cup \{x\}$ is an independent subset of $A \cup \{x\}$. Therefore $\rho(A \cup \{x\}) > \rho(A)$; and A is a flat. $\qquad\square$

The intersection of all flats containing a given subset A of E is called the *span* of A. By Theorem 2.7, the span is itself a flat, in fact the smallest flat containing A. We shall see in the exercises that the span of A is the set $\{x \in E : \rho(A \cup \{x\}) = \rho(A)\}$.

We end this section with an example to illustrate all the concepts introduced above.

Example

Let $E = \{1, 2, \ldots, 10\}$ and let $\mathscr{E}$ consist of those subsets of E of three or fewer members whose corresponding sets of lines in the figure include no triangle. It is not difficult to check that $(E, \mathscr{E})$ is an independence space. (It is closely related to the graphic spaces which we shall meet in Chapter 3.) To illustrate I(2), we note that $\{1, 6\}$,

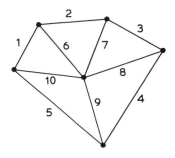

$\{5, 6, 10\} \in \mathscr{E}$ and from these sets also $\{1, 5, 6\} \in \mathscr{E}$. Next, $\{1, 6, 7\}$, $\{2, 3, 7\}$ are bases, $1 \in \{1, 6, 7\}$ and $(\{1, 6, 7\} \backslash \{1\}) \cup \{3\}$ is also a basis; which illustrates B(2). To illustrate C(2)', we note that $\{1, 2, 3, 7\}$, $\{1, 6, 10\}$ are circuits and $\{2, 3, 6, 7, 10\}$ contains, for instance, the circuits $\{2, 3, 6, 10\}$ and $\{2, 3, 7, 10\}$. Finally, $\{2, 6\}$ has rank 2, and the largest set containing $\{2, 6\}$ which is also of rank 2 is $\{2, 6, 7\}$; so this is a flat which is the span of $\{2, 6\}$. Observe that the flats $\{2, 6, 7\}$ and $\{1, 6, 10\}$ provide an example which shows that, of course, a union of flats need not itself be a flat.

2.2 Some induced structures

Given an independence structure (or several structures) there are various ways in which it generates or 'induces' others. Throughout this section we let $(E, \mathscr{E})$ be an independence space with rank function ρ and we suppose that $E' \subseteq E$. Whenever we use the word 'independent' without further qualification, we mean 'independent in $(E, \mathscr{E})$'.

One of the simplest induced structures is that of restriction. As we have already noted in Section 2.1, the collection $\mathscr{E} | E'$ of members of $\mathscr{E}$ which are subsets of E' is an independence structure on E', called the restriction of $\mathscr{E}$ to E'; its rank function is, of course, just the rank function of $\mathscr{E}$ restricted to the subsets of E'. Somewhat akin to restriction is 'contraction'. Let us define the collection $\mathscr{E}_{\otimes E \backslash E'}$ of subsets of $E \backslash E'$ by the condition:

$$A \in \mathscr{E}_{\otimes E \backslash E'} \Leftrightarrow A \cup B \in \mathscr{E} \text{ for some basis } B \text{ of } \mathscr{E} | E'.$$

Then $\mathscr{E}_{\otimes E \backslash E'}$ is called the *contraction of $\mathscr{E}$ away from E'*.

Theorem 2.8. *$\mathscr{E}_{\otimes E \backslash E'}$ is an independence structure on $E \backslash E'$, and its rank function $\rho_{\otimes}$ is given by the formula*:

$$\rho_{\otimes}(A) = \rho(A \cup E') - \rho(E') \quad \forall A \subseteq E \backslash E'.$$

Proof. If $A \in \mathscr{E}_{\otimes E \backslash E'}$, then $A \cup B \in \mathscr{E}$ for some basis B of $\mathscr{E} | E'$ and so $A' \cup B \in \mathscr{E}$ for any $A' \subseteq A$. Hence $\mathscr{E}_{\otimes E \backslash E'}$ trivially satisfies I(1). To verify I(2), let $A, A' \in \mathscr{E}_{\otimes E \backslash E'}$ with $|A'| = |A| + 1$, and let B, B' be bases of $\mathscr{E} | E'$ such that $A \cup B, A' \cup B' \in \mathscr{E}$. By I(2) applied to B' and $A \cup B$ in $\mathscr{E}$, there exists $X \in \mathscr{E}$ with $|X| = |A \cup B| (= |A \cup B'|)$ and $B' \subseteq X \subseteq (A \cup B) \cup B'$; and so it follows that $X = A \cup B'$. Hence

$A \cup B'$, $A' \cup B' \in \mathscr{E}$ and $|A' \cup B'| = |A \cup B'| + 1$; and, again by I(2) in $\mathscr{E}$, there exists $y \in (A' \cup B') \backslash (A \cup B') = A' \backslash A$ such that $A \cup \{y\} \cup B' \in \mathscr{E}$. Thus $A \cup \{y\} \in \mathscr{E}_{\otimes E \backslash E'}$; and I(2) is verified for this structure.

Now let $A \subseteq E \backslash E'$ and let A' be a maximal independent subset of A in $\mathscr{E}_{\otimes E \backslash E'}$; then $\rho_\otimes(A) = |A'|$ and $A' \cup B \in \mathscr{E}$ for some basis B of $\mathscr{E}|E'$ (in fact, as above, for *any* basis B of $\mathscr{E}|E'$). Further, $A' \cup B \subseteq A \cup E'$ and clearly is a maximal independent subset of $A \cup E'$. Therefore

$$\rho(A \cup E') = |A' \cup B| = |A'| + |B| = \rho_\otimes(A) + \rho(E');$$

and the required link between ρ and $\rho_\otimes$ is established. $\qquad\square$

Example

Let $E = \{1, 2, \ldots, 8\}$ and let $\mathscr{E}$ be the independence structure with bases consisting of those subsets of E of four elements whose corresponding points in the tetrahedron are not coplanar.

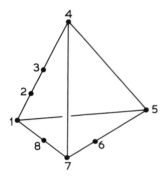

'Independence' here is just affine independence in real three-dimensional space. Let $E' = \{1, 2, 3,\}$. Then clearly

$$\mathscr{E}|E' = \{\phi, \{1\}, \{2\}, \{3\}, \{1, 2\}, \{1, 3\}, \{2, 3\}\}.$$

Furthermore, $\{5, 6\} \in \mathscr{E}_{\otimes E \backslash E'}$ since, for example, $\{1, 2\}$ is a basis of $\mathscr{E}|E'$ such that $\{1, 2, 5, 6\} \in \mathscr{E}$. In fact, it is easy to see that the bases of $\mathscr{E}_{\otimes E \backslash E'}$ are just $\{5, 6\}$, $\{5, 7\}$, $\{5, 8\}$, $\{6, 7\}$ and $\{6, 8\}$.

Now let $(E_1, \mathscr{E}_1)$ and $(E_2, \mathscr{E}_2)$ be independence spaces with $E_1 \cap E_2 = \phi$. The collection of those subsets of $E_1 \cup E_2$ which are unions of members of $\mathscr{E}_1$ and $\mathscr{E}_2$ is called the *direct sum* of $\mathscr{E}_1$ and $\mathscr{E}_2$

and is denoted by $\mathscr{E}_1 \oplus \mathscr{E}_2$. The verification that it is an independence structure is trivial. If $A \subseteq E_1 \cup E_2$, then its rank in $\mathscr{E}_1 \oplus \mathscr{E}_2$ is evidently equal to $\rho_1(A \cap E_1) + \rho_2(A \cap E_2)$, where ρ_1, ρ_2 are the respective rank functions of $\mathscr{E}_1, \mathscr{E}_2$. We may, of course, also regard each $\mathscr{E}_i$ as an independence structure on $E_1 \cup E_2$, and then we may rewrite the rank of A in their direct sum as $\rho_1(A) + \rho_2(A)$. Obviously this notion extends to any number of structures on pairwise-disjoint sets. The notion of a direct sum is related in Section 2.4 below to the much more significant notion of a general sum.

Now let us define the collection $\mathscr{E}(E')$ of subsets of E by the condition

$$A \in \mathscr{E}(E') \Leftrightarrow A \cup B \in \mathscr{E} \text{ for some basis } B \text{ of } \mathscr{E}|E'.$$

At first glance this seems to be the same as the contraction $\mathscr{E}_{\otimes E \backslash E'}$; but we note that no longer are we assuming that $A \subseteq E \backslash E'$, and hence the set B in the definition may not be disjoint from A. We observe that $\mathscr{E}(E')$ is an independence structure on E closely related, of course, to $\mathscr{E}_{\otimes E \backslash E'}$.

Theorem 2.9. $\mathscr{E}(E')$ *is an independence structure on* E. *Indeed, it is the direct sum of* $\mathscr{E}|E'$ *and* $\mathscr{E}_{\otimes E \backslash E'}$, *and so its rank function* $\rho_{E'}$ *is given by*

$$\rho_{E'}(A) = \rho(A \cup E') + \rho(A \cap E') - \rho(E') \quad \forall A \subseteq E. \qquad \square$$

We now turn to quite a different induced structure. Let us define the collection $\mathscr{E}^*$ of subsets of E by the condition

$$A \in \mathscr{E}^* \Leftrightarrow A \cap B = \phi \text{ for some basis } B \text{ of } E.$$

Then $\mathscr{E}^*$ is called the *dual* of $\mathscr{E}$.

Theorem 2.10. $\mathscr{E}^*$ *is an independence structure on* E, *and its rank function* ρ^* *is given by the formula*

$$\rho^*(A) = |A| - \rho(E) + \rho(E \backslash A) \quad \forall A \subseteq E.$$

Further $(\mathscr{E}^*)^* = \mathscr{E}$.

Proof. Instead of giving a direct proof (which is reserved for an exercise), we invoke Theorem 2.6. Let, then, $\mathscr{C}^*$ be the collection of (certainly not empty) minimal subsets of E which are not members of $\mathscr{E}^*$. Then $X \in \mathscr{E}^*$ if and only if X contains no member of $\mathscr{C}^*$ as a subset. We verify that $\mathscr{C}^*$ satisfies the circuit axioms. Certainly, by their minimality, the members of $\mathscr{C}^*$ satisfy C(1). Now assume that $\mathscr{C}^*$ fails to satisfy C(2)'. Then there exist $C, C' \in \mathscr{C}^*$ with $C \neq C'$ and $x \in C \cap C'$ such that $(C \cup C') \backslash \{x\}$ does not contain a member of $\mathscr{C}^*$. But then $(C \cup C') \backslash \{x\} \in \mathscr{E}^*$, and there exists a basis B of $\mathscr{E}$ with $((C \cup C') \backslash \{x\}) \cap B = \phi$. Since $C \cap B$ and $C' \cap B$ are non-empty, it follows that $C \cap B = C' \cap B = \{x\}$. Now, as $C \nsubseteq C'$, there exists $y \in C \backslash C'$ and, by the minimality of C, it follows that $C \backslash \{y\} \in \mathscr{E}^*$. Therefore $(C \backslash \{y\}) \cap B' = \phi$ for some basis B' of $\mathscr{E}$. By B(2), applied in $\mathscr{E}$, there exists $z \in B'$ $(\subseteq (E \backslash C) \cup \{y\})$ such that $B'' = (B \backslash \{x\}) \cup \{z\}$ is a basis of $\mathscr{E}$. Either $z = y$, in which case $z \notin C'$ and $B'' \cap C' = \phi$; or $z \neq y$, in which case $z \in E \backslash C$ and $B'' \cap C = \phi$. Each of these situations is impossible; and so this contradiction shows that C(2)' is satisfied by $\mathscr{C}^*$. Hence, by Theorem 2.6, $\mathscr{E}^*$ is an independence structure on E.

Before establishing the formula for the rank function, we observe that the maximal members of $\mathscr{E}^*$, or its bases, are precisely the complements in E of the bases of $\mathscr{E}$. Therefore the relation between $\mathscr{E}$ and $\mathscr{E}^*$ is a symmetrical one and so $(\mathscr{E}^*)^* = \mathscr{E}$, and also $\rho^*(E) + \rho(E) = |E|$.

Now let $A \subseteq E$, let A' be a maximal independent subset of $E \backslash A$, and let B be a basis of $\mathscr{E}$ with $A' \subseteq B$. Then clearly $B \backslash A' \subseteq A$ (by the maximality of A' in $E \backslash A$) and $A \backslash B \in \mathscr{E}^*$. Therefore

$$\rho^*(A) \geqslant |A \backslash B| = |A| - |A \cap B| = |A| - |B \backslash A'|$$
$$= |A| - |B| + |A'|$$
$$= |A| - \rho(E) + \rho(E \backslash A).$$

Finally, since $(\mathscr{E}^*)^* = \mathscr{E}$, the replacement of ρ by ρ^* and A by $E \backslash A$ throughout this inequality gives

$$\rho(E \backslash A) \geqslant |E \backslash A| - \rho^*(E) + \rho^*(A)$$
$$= |E| - |A| - \rho^*(E) + \rho^*(A)$$
$$= \rho(E) - |A| + \rho^*(A).$$

The two above inequalities together give the desired result

$$\rho^*(A) = |A| - \rho(E) + \rho(E \backslash A). \qquad \square$$

Example

Let $E = \{1, 2, 3, 4, 5, 6, 7\}$ and let the bases of $\mathscr{E}$ be those subsets of E

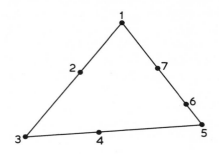

of three elements whose corresponding points in the figure are not collinear. 'Independence' here is just affine independence in the ordinary real plane. Let $E' = \{1, 2, 3\}$. Then, for example, $\{1, 2, 4\} \in \mathscr{E}(E')$ since $\{1, 2\}$ is a basis of $\mathscr{E}|E'$ and $\{1, 2, 4\} \in \mathscr{E}$. In fact, $\mathscr{E}(E')$ has bases $\{x, y, z\}_{\neq}$, where $x, y \in \{1, 2, 3\}$ and $z \in \{4, 5, 6, 7\}$. Also, for example, $\mathscr{E}*$ has a basis $\{1, 2, 3, 5\}$, since $\{4, 6, 7\}$ is a basis of $\mathscr{E}$. In fact, the bases of $\mathscr{E}*$ are all sets of four elements of E apart from $\{4, 5, 6, 7\}$, $\{1, 2, 6, 7\}$, $\{1, 2, 3, 4\}$, $\{2, 3, 4, 5\}$, $\{2, 3, 4, 6\}$ and $\{2, 3, 4, 7\}$.

We now digress briefly from our theme of induced structures. Let us associate with each $x \in E$ a non-negative real number called its *weight*, $w(x)$ say. For $A \subseteq E$, we shall write $w(A) = \sum_{x \in A} w(x)$ and call this also the weight of A. The theorem below describes an inductive procedure for finding a basis of $(E, \mathscr{E})$ of maximum weight. It is the most naive method, and the surprise is that it actually works: it is known as the *'greedy algorithm'*.

Theorem 2.11. *A heaviest basis of $(E, \mathscr{E})$ (i.e. one of maximum weight) can be found by the following inductive procedure*

I. *Put $A_0 = \phi$.*
II. *Suppose that $0 \leqslant i < r = \rho(E)$ and that sets $A_0, A_1, \ldots, A_i$ have been constructed. Let A_{i+1} be formed by adjoining to A_i an element $x_{i+1} \in E \backslash A_i$ with $w(x_{i+1})$ as large as possible subject to the condition that $A_i \cup \{x_{i+1}\} \in \mathscr{E}$.*
Then A_r is a heaviest basis.

Proof. Certainly $A_r = \{x_1, \ldots, x_r\}$ is a basis. Assume, then, that the procedure fails to produce a heaviest basis and that $B = \{y_1, \ldots, y_r\}$

is a basis with $w(A_r) < w(B)$ and, say, $w(y_1) \geq w(y_2) \geq \ldots \geq w(y_r)$. Then there must exist j with $1 \leq j \leq r$ such that $w(x_j) < w(y_j)$. Consider $\{x_1, \ldots, x_{j-1}\}$, $\{y_1, \ldots, y_j\}$ $(\in \mathscr{E})$. By the replacement property I(2), there exists k with $1 \leq k \leq j$ such that $\{x_1, \ldots, x_{j-1}, y_k\} \neq \in \mathscr{E}$. But then $y_k \in E \backslash \{x_1, \ldots, x_{j-1}\} = E \backslash A_{j-1}$ and

$$w(y_k) \geq w(y_j) > w(x_j);$$

which contradicts the choice of x_j in the construction of A_j. Hence the procedure does, in fact, produce a heaviest basis.   □

An interpretation of this simple theorem for graphs will be described in the next chapter. Now, however, we use the concept of heaviest bases to describe another induced structure. Note before proceeding that, if B and B' are bases of an independence space and $x \in B$, then there exists $y \in B'$ such that *both* of $(B \backslash \{x\}) \cup \{y\}$ and $(B' \backslash \{y\}) \cup \{x\}$ are bases of the space. The proof of this generalization of B(2) is left as an exercise at the end of the chapter.

Theorem 2.12. *Let $\mathscr{B}'$ be the collection of heaviest bases of $(E, \mathscr{E})$ (with respect to some given weighting). Then $\mathscr{B}'$ is the collection of bases of an independence structure on E.*

Proof. Denote the weighting considered by w. Since the collection of all the bases of $(E, \mathscr{E})$ satisfies B(1), it is at once clear that so also does the subcollection $\mathscr{B}'$. To show that $\mathscr{B}'$ satisfies B(2), let $B, B' \in \mathscr{B}'$ and let $x \in B$. Then B and B' are bases of $\mathscr{E}$ and so, by the comments immediately preceding the theorem, there exists $y \in B'$ such that $(B \backslash \{x\}) \cup \{y\}$ and $(B' \backslash \{y\}) \cup \{x\}$ are both bases of $\mathscr{E}$. Since B' is a heaviest basis, it follows that $w((B' \backslash \{y\}) \cup \{x\}) \leq w(B')$ and hence that $w(x) \leq w(y)$. Therefore $w(B) \leq w((B \backslash \{x\}) \cup \{y\})$, and it follows that B and $(B \backslash \{x\}) \cup \{y\}$ are both heaviest bases. Thus $(B \backslash \{x\}) \cup \{y\} \in \mathscr{B}'$; and we have verified that $\mathscr{B}'$ satisfies B(1) and B(2). Hence, by Theorem 2.5, $\mathscr{B}'$ is the collection of bases of an independence structure on E.   □

A special case of this last induced structure is worth a moment's examination. Let $E' \subseteq E$, and define a weighting w by the rule

$$w(x) = \begin{cases} 1 & \text{if} \quad x \in E' \\ 0 & \text{if} \quad x \notin E' \end{cases}$$

Then the heaviest bases of $(E, \mathscr{E})$ with respect to this particularly simple weighting of the elements of E are precisely those bases of $(E, \mathscr{E})$ which contain a maximal independent subset of E', and these are just the bases of $\mathscr{E}(E')$. So here we have an alternative proof that $\mathscr{E}(E')$ (and hence $\mathscr{E}_{\otimes E \setminus E'}$) is an independence structure.

We conclude by mentioning yet another, particularly simple, induced structure. Let the rank of $\mathscr{E}$ be r, and let s be an integer with $0 \leqslant s \leqslant r$. Those members of $\mathscr{E}$ consisting of s or fewer elements form another independence structure on E, known as the *truncation* of $\mathscr{E}$ at s; and a *proper truncation* of $\mathscr{E}$ if $s < r$. The simple verification of this fact is left as an exercise at the end of the chapter.

We have spoken throughout of induced 'structures', but we shall feel free to speak in similar vein of spaces: thus of contracted, dual, truncated spaces, etc.

2.3 Submodular functions

We have already described some simple methods of constructing new independence spaces from a given independence space; for example, its dual, its restrictions and contractions, and the space arising from its heaviest bases. In this section we describe a very general method for the construction of independence spaces based on a generalization of the rank function. So, to start our discussion, we prove a result, postponed from Section 2.1, that the properties R(1), R(2) and R(3) of a rank function are characteristic properties.

Theorem 2.13. *Let E be a given set, and let $\rho : \mathscr{P}(E) \to \{0, 1, 2, \ldots\}$ satisfy, for each $A, B \subseteq E$, the properties*

R(1) $\rho(A) \leqslant |A|$;
R(2) *if $B \subseteq A$, then $\rho(B) \leqslant \rho(A)$;*
R(3) $\rho(A) + \rho(B) \geqslant \rho(A \cup B) + \rho(A \cap B)$.

Then the collection $\mathscr{E}$ of subsets A of E for which $\rho(A) = |A|$ forms an independence structure on E, and ρ is its rank function.

Proof. Note first that $0 \leqslant \rho(\phi) \leqslant |\phi| = 0$ and so $\phi \in \mathscr{E}$, and $\mathscr{E}$ is nonempty. We shall now show that $\mathscr{E}$ satisfies I(1) and I(2). To verify I(1), we observe that, if $A \in \mathscr{E}$ and $B \subseteq A$, then

$$|B| + |A \backslash B| = |A| = \rho(A) = \rho(B \cup (A \backslash B)) + \rho(B \cap (A \backslash B))$$
$$\leqslant \rho(B) + \rho(A \backslash B) \leqslant |B| + |A \backslash B|.$$

Hence there is equality throughout and, in particular, $\rho(B) = |B|$ and so $B \in \mathscr{E}$.

Before proceeding to the proof of I(2), we claim that, if $B \subseteq E$, then any maximal subset of B in $\mathscr{E}$ has cardinality equal to $\rho(B)$. For let us assume otherwise, let C be a maximal subset of B in $\mathscr{E}$ with $|C| = \rho(C) < \rho(B)$, and let D be maximal with respect to the properties $C \subseteq D \subseteq B$ and $\rho(C) = \rho(D)$. Then $D \neq B$, and so there exists $y \in B \backslash D$. It then follows that

$$\rho(D) + \rho(C \cup \{y\}) \geqslant \rho(D \cup C \cup \{y\}) + \rho(D \cap (C \cup \{y\}))$$
$$= \rho(D \cup \{y\}) + \rho(C)$$
$$> \rho(D) + \rho(C)$$

by the maximality of D. Hence

$$|C| + 1 = |C \cup \{y\}| \geqslant \rho(C \cup \{y\}) > \rho(C) = |C|,$$

so $|C \cup \{y\}| = \rho(C \cup \{y\})$, and $C \cup \{y\} \in \mathscr{E}$; which contradicts the maximality of $C \ (\subseteq B)$ in $\mathscr{E}$. Therefore, as claimed, any maximal subset of B in $\mathscr{E}$ has cardinality $\rho(B)$.

From this fact we are now easily able to verify I(2) and to complete the proof. For, if $A, A' \in \mathscr{E}$ and $|A| < |A'|$, then

$$\rho(A) = |A| < |A'| = \rho(A') \leqslant \rho(A \cup A').$$

Hence, by the above argument, A is not a maximal subset of $A \cup A'$ in $\mathscr{E}$ and so there exists $y \in (A \cup A') \backslash A = A' \backslash A$ with $A \cup \{y\} \in \mathscr{E}$; thus I(2) is verified. Therefore $\mathscr{E}$ is an independence structure. Also, for each $B \subseteq E$, any maximal independent subset of B has cardinality $\rho(B)$; in other words, ρ is the rank function of $\mathscr{E}$. □

As a very simple example, we observe that the function $\rho : \mathscr{P}(E) \rightarrow \{0, 1, 2, \ldots\}$ defined by the condition $\rho(A) = |A|$ for each $A \subseteq E$ clearly satisfies R(1), R(2) and R(3), and so is the rank function of an independence structure (indeed, of the universal structure on E).

In the following definition we now replace property R(1) by the weaker property R(0) (below) in order to introduce the promised generalization of the rank function. We then see from our next theorem exactly how these more general functions are related to independence spaces.

A function $\mu : \mathscr{P}(E) \to \{0, 1, 2, \ldots\}$ with the properties

R(0) $\mu(\phi) = 0$;
R(2) if $B \subseteq A$, then $\mu(B) \leqslant \mu(A)$;
R(3) $\mu(A) + \mu(B) \geqslant \mu(A \cup B) + \mu(A \cap B)$

for all $A, B \subseteq E$, is called *submodular* on E.

Rank functions are, of course, necessarily submodular. On the other hand, not every submodular function is a rank function; one such appears in the proof of Theorem 2.16 below.

Theorem 2.14. *Let μ be a submodular function on a given set E, and let $\mathscr{E}$ be the collection of subsets of E defined by the condition*

$$A \in \mathscr{E} \Leftrightarrow \mu(B) \geqslant |B| \quad \forall B \subseteq A.$$

Then $\mathscr{E}$ is an independence structure on E, and its rank function ρ is given by the formula

$$\rho(A) = \min_{B \subseteq A} \{\mu(B) + |A \backslash B|\} \quad \forall A \subseteq E.$$

Proof. Our proof is in two parts: we shall show first that the given function ρ satisfies R(1), R(2) and R(3) (and hence, by Theorem 2.13, that it is the rank function of an independence structure on E), and second that $\rho(A) = |A|$ if and only if $A \in \mathscr{E}$ (and hence, again by Theorem 2.13, that $\mathscr{E}$ is precisely the independence structure whose rank function is ρ).

First, then, let us observe that ρ is certainly defined for all subsets of E and takes non-negative integer values. Also, for $A \subseteq E$,

$$\rho(A) = \min_{B \subseteq A} \{\mu(B) + |A \backslash B|\} \leqslant \mu(\phi) + |A \backslash \phi| = |A|;$$

and so R(1) is established. Now let $B \subseteq A$. Then there exists $C \subseteq A$ such that $\rho(A) = \mu(C) + |A \backslash C|$. If we write $C' = B \cap C$, then

$$\rho(A) = \mu(C) + |A \backslash C| \geqslant \mu(B \cap C) + |B \backslash C|$$
$$= \mu(C') + |B \backslash C'| \geqslant \rho(B);$$

which establishes R(2). Similarly, for any A, $B \subseteq E$, there exist $C \subseteq A$, $D \subseteq B$ such that $\rho(A) = \mu(C) + |A \backslash C|$ and $\rho(B) = \mu(D) + |B \backslash D|$. Now it is easy to verify that

$$|A \backslash C| + |B \backslash D| = |(A \cup B) \backslash (C \cup D)| + |(A \cap B) \backslash (C \cap D)|$$

and hence, since μ is submodular,

$$\begin{aligned} \rho(A) + \rho(B) = \mu(C) &+ |A \backslash C| + \mu(D) + |B \backslash D| \\ &\geqslant \mu(C \cup D) + \mu(C \cap D) + |(A \cup B) \backslash (C \cup D)| \\ &\quad + |(A \cap B) \backslash (C \cap D)| \\ &\geqslant \rho(A \cup B) + \rho(A \cap B). \end{aligned}$$

Hence R(3) is established; and ρ is indeed the rank function of an independence space.

Finally, for any $A \subseteq E$,

$$\rho(A) = |A| \Rightarrow \mu(B) + |A \backslash B| \geqslant |A| \quad \forall B \subseteq A$$
$$\Rightarrow \mu(B) \geqslant |B| \quad \forall B \subseteq A$$

and

$$\mu(B) \geqslant |B| \quad \forall B \subseteq A \Rightarrow \rho(A) = \min_{B \subseteq A} \{\mu(B) + |A \backslash B|\}$$
$$\geqslant \min_{B \subseteq A} \{|B| + |A \backslash B|\} = |A| \geqslant \rho(A)$$
$$\Rightarrow \rho(A) = |A|;$$

i.e. $\qquad \mu(B) \geqslant |B| \quad \forall B \subseteq A \Leftrightarrow \rho(A) = |A|.$

Hence $A \in \mathscr{E}$ if and only if $\rho(A) = |A|$; and therefore $\mathscr{E}$ is an independence structure with rank function ρ as required. $\qquad \square$

We shall now apply the theory of submodular functions which we have just developed to show how a given bipartite graph and an independence space together induce a new independence space in quite a striking fashion. Our discussion has close connections with the kind of independence spaces introduced later in Chapter 4. The reader may recall the notion of a bipartite graph from Section 1.2 and also that of a matching, which we shall have occasion to use. It is now convenient to adopt the notation $G = (E, \Delta, E')$ for the particular bipartite graph under consideration; here E and E' are the two non-empty disjoint subsets into which the vertex set of G is partitioned (though we have used the letter E, they are *not* sets of edges), and Δ is the set of edges of G. If $A \subseteq E$, we shall write

$$\Delta(A) = \{e \in E' : ae \in \Delta \text{ for some } a \in A\};$$

with a similar convention for subsets of E'. Instead of $\Delta(\{x\})$, we shall write simply $\Delta(x)$. Observe that a matching in G is, of course, a particular subset of Δ. If $A(\subseteq E)$ and $B(\subseteq E')$ are precisely the endpoints of a matching in G, we shall say that 'A and B are matched', that there exists 'a matching between A and B', or some other similar expression. A matching may be an empty set of edges.

Before we prove our main theorem (Theorem 2.16), we shall need a preliminary result, very important in its own right. First, we illustrate Theorem 2.16 by an example.

Example

Let $G = (E, \Delta, E')$ be the bipartite graph illustrated and $\mathscr{E}'$ the independence structure on E' consisting of all subsets of E' of cardinality three or less with the exception of $\{1, 2, 3\}$. We observe that there is clearly a matching in G with respect to which $\{a, b\}$ and $\{1, 2\}$ are matched. However, $\{a, b, c\}$ is never matched in G with any of the members of $\mathscr{E}'$. In fact, the collection of subsets of E matched with the

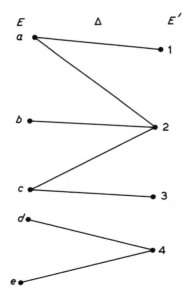

subsets of E' belonging to $\mathscr{E}'$ is precisely

$$\mathscr{E} = \{\phi, \{a\}, \{b\}, \{c\}, \{d\}, \{e\}, \{a, b\}, \{a, c\}, \{a, d\}, \{a, e\},$$
$$\{b, c\}, \{b, d\}, \{b, e\}, \{c, d\}, \{c, e\}, \{a, b, d\}, \{a, b, e\},$$
$$\{a, c, d\}, \{a, c, e\}, \{b, c, d\}, \{b, c, e\}\};$$

and $\mathscr{E}$ is itself an independence structure on E.

Theorem 2.15. *Let* $G = (E, \Delta, E')$ *be a bipartite graph and* $\mathscr{E}'$ *an independence structure on* E' *with rank function* ρ'. *Then, for* $A \subseteq E$, *there exists a matching in* G *between* A *and an independent subset of* E' *if and only if*

$$\rho'(\Delta(B)) \geqslant |B| \quad \forall B \subseteq A.$$

Proof. If there exists such a matching, say $\Delta^* \, (\subseteq \Delta)$, then, for each $B \subseteq A$, we have

$$\rho'(\Delta(B)) \geqslant \rho'(\Delta^*(B)) = |\Delta^*(B)| = |B|,$$

since $\Delta^*(B) \subseteq \Delta^*(A) \in \mathscr{E}'$.

Conversely, let us assume that $A \subseteq E$ is such that $\rho'(\Delta(B)) \geqslant |B|$ for each $B \subseteq A$, and let $\Delta^* \, (\subseteq \Delta)$ be minimal with respect to the property that $\rho'(\Delta^*(B)) \geqslant |B|$ for each $B \subseteq A$. We shall show that Δ^* is the required matching. For certainly, for each $x \in A$, $\rho'(\Delta^*(x)) \geqslant |\{x\}| = 1$, and so $\Delta^*(x) \neq \phi$. Assume that $|\Delta^*(x)| > 1$ for some $x \in A$, let δ_1, δ_2 be different numbers of Δ^* with endpoint x, and let $\Delta_1 = \Delta^* \backslash \{\delta_1\}$, $\Delta_2 = \Delta^* \backslash \{\delta_2\}$. Then, by the minimality of Δ^*, there exist $C_1, C_2 \subseteq A$ with $\rho'(\Delta_1(C_1)) \leqslant |C_1| - 1$ and $\rho'(\Delta_2(C_2)) \leqslant |C_2| - 1$. Also, since $\rho'(\Delta^*(C_1)) \geqslant |C_1|$, it is clear that $x \in C_1$; similarly $x \in C_2$. Note also that

$$\Delta_1(C_1) \cup \Delta_2(C_2) = \Delta^*(C_1) \cup \Delta^*(C_2) \, (= \Delta^*(C_1 \cup C_2))$$

and that

$$\Delta_1(C_1) \cap \Delta_2(C_2) \supseteq \Delta^*(C_1 \backslash \{x\}) \cap \Delta^*(C_2 \backslash \{x\}) \supseteq \Delta^*((C_1 \cap C_2) \backslash \{x\}).$$

Hence

$$(|C_1| - 1) + (|C_2| - 1) \geqslant \rho'(\Delta_1(C_1)) + \rho'(\Delta_2(C_2))$$
$$\geqslant \rho'(\Delta_1(C_1) \cup \Delta_2(C_2)) + \rho'(\Delta_1(C_1) \cap \Delta_2(C_2))$$
$$\geqslant \rho'(\Delta^*(C_1 \cup C_2)) + \rho'(\Delta^*((C_1 \cup C_2) \backslash \{x\}))$$
$$\geqslant |C_1 \cup C_2| + |(C_1 \cap C_2) \backslash \{x\}|$$
$$= |C_1| + |C_2| - 1;$$

which is manifestly a contradiction. Hence $|\Delta^*(x)| = 1$ for each $x \in A$. Thus $|A| \geq |\Delta^*(A)| \geq \rho'(\Delta^*(A)) \geq |A|$, and so clearly there is equality throughout; and it follows that Δ^* is a matching between A and $\Delta^*(A) \in \mathscr{E}'$. $\qquad\square$

Theorem 2.16. *Let* $G = (E, \Delta, E')$ *be a bipartite graph and* $\mathscr{E}'$ *an independence structure on* E' *with rank function* ρ'. *Let* $\mathscr{E}$ *denote the collection of subsets of* E *each of which can be matched in* G *with an independent subset of* E'. *Then* $\mathscr{E}$ *is an independence structure with rank function* ρ *given by the formula*

$$\rho(A) = \min_{B \subseteq A} \{\rho'(\Delta(B)) + |A \setminus B|\} \quad \forall A \subseteq E.$$

Proof. We note first that, by Theorem 2.15, $A \in \mathscr{E}$ if and only if $\rho'(\Delta(B)) \geq |B|$ for each $B \subseteq A$. A reference to Theorem 2.14 will now give the key to the remainder of this proof. Indeed, let $\mu(A) = \rho'(\Delta(A))$ for each $A \subseteq E$; we proceed to show that μ is a submodular function on E. The properties R(0) and R(2) are immediate and, for $A, B \subseteq E$,

$$\begin{aligned}
\mu(A) + \mu(B) &= \rho'(\Delta(A)) + \rho'(\Delta(B)) \\
&\geq \rho'(\Delta(A) \cup \Delta(B)) + \rho'(\Delta(A) \cap \Delta(B)) \\
&\geq \rho'(\Delta(A \cup B)) + \rho'(\Delta(A \cap B)) \\
&= \mu(A \cup B) + \mu(A \cap B);
\end{aligned}$$

and we conclude that R(3) also holds for μ. Thus μ is submodular and, by Theorem 2.14, $\mathscr{E}$ is an independence structure whose rank function ρ is given by

$$\rho(A) = \min_{B \subseteq A} \{\mu(B) + |A \setminus B|\};$$

which is exactly as required. $\qquad\square$

2.4 Sums of independence structures

This section is devoted to yet another induced structure closely related to, but much more interesting than, that of a direct sum. Let, then, $\mathscr{E}_1$ and $\mathscr{E}_2$ be independence structures on a set E with respective rank functions ρ_1 and ρ_2. Then the collection $\mathscr{E}$ of subsets of E defined by the condition:

$$A \in \mathscr{E} \Leftrightarrow A = A_1 \cup A_2 \text{ for some } A_1 \in \mathscr{E}_1 \text{ and } A_2 \in \mathscr{E}_2,$$

is called the *sum* of $\mathscr{E}_1$ and $\mathscr{E}_2$ (written $\mathscr{E}_1 + \mathscr{E}_2$), and turns out itself to be an independence structure. However, the reader who tries to establish this fact from basic principles will find it quite hard work. We have postponed studying sums until now in order to be able to make use of the theory of independence structures induced by bipartite graphs as discussed in Section 2.3. We note before proceeding that since $\mathscr{E}_1$ and $\mathscr{E}_2$ satisfy the hereditary property I(1), any member of $\mathscr{E}_1 + \mathscr{E}_2$ can be written as $A_1 \cup A_2$ for some *disjoint* $A_1 \in \mathscr{E}_1$ and $A_2 \in \mathscr{E}_2$.

Theorem 2.17. *Let $\mathscr{E}_1$, $\mathscr{E}_2$ be independence structures on E with rank functions ρ_1, ρ_2 respectively. Then their sum $\mathscr{E} = \mathscr{E}_1 + \mathscr{E}_2$ is an independence structure on E whose rank function ρ is given by the formula:*

$$\rho(A) = \min_{B \subseteq A} \{\rho_1(B) + \rho_2(B) + |A \setminus B|\} \quad \forall A \subseteq E.$$

Proof. Write $E = \{x_1, \ldots, x_n\}_{\neq}$, let $x_1, \ldots, x_n, y_1, \ldots, y_n, z_1, \ldots, z_n$ be distinct, and write $E_1 = \{y_1, \ldots, y_n\}$, $E_2 = \{z_1, \ldots, z_n\}$. We define

$$\mathscr{E}'_1 = \{\{y_{i_1}, \ldots, y_{i_j}\} : \{i_1, \ldots, i_j\}_{\neq} \subseteq \{1, \ldots, n\} \text{ and } \{x_{i_1}, \ldots, x_{i_j}\} \in \mathscr{E}_1\}$$

and

$$\mathscr{E}'_2 = \{\{z_{i_1}, \ldots, z_{i_j}\} : \{i_1, \ldots, i_j\}_{\neq} \subseteq \{1, \ldots, n\} \text{ and } \{x_{i_1}, \ldots, x_{i_j}\} \in \mathscr{E}_2\}.$$

Then $\mathscr{E}'_1$ and $\mathscr{E}'_2$ are just copies of $\mathscr{E}_1$ and $\mathscr{E}_2$ on $E_1 \cup E_2 (= E'$ say), and

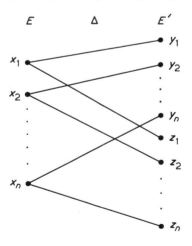

so they are, of course, independence structures themselves. Let their rank functions be ρ_1' and ρ_2'. Since $E_1 \cap E_2 = \phi$, the direct sum $\mathscr{E}' = \mathscr{E}_1' \oplus \mathscr{E}_2'$ may be defined on E' and, as in Section 2.2, its rank function ρ' is simply given by $\rho_1' + \rho_2'$.

Now let us consider the bipartite graph $G = (E, \Delta, E')$, where $\Delta = \{x_1 y_1, \ldots, x_n y_n, x_1 z_1, \ldots, x_n z_n\}$ as illustrated above. By Theorem 2.16, the collection of subsets A of E each of which can be matched in G with a member of $\mathscr{E}'$ is itself on independence structure on E. But this collection is readily seen to be $\mathscr{E}$ itself; and so the first assertion of the theorem is established.

Finally, we observe that, for each $B \subseteq E$, say $B = \{x_1, \ldots, x_m\}$,

$$\rho_1'(\Delta(B)) = \rho_1'(\{y_1, \ldots, y_m, z_1, \ldots, z_m\})$$
$$= \rho_1'(\{y_1, \ldots, y_m\}) = \rho_1(B);$$

and a similar formula holds for ρ_2'. Hence, again by Theorem 2.16,

$$\rho(A) = \min_{B \subseteq A} \{\rho'(\Delta(B)) + |A \backslash B|\}$$

$$= \min_{B \subseteq A} \{\rho_1'(\Delta(B)) + \rho_2'(\Delta(B)) + |A \backslash B|\}$$

$$= \min_{B \subseteq A} \{\rho_1(B) + \rho_2(B) + |A \backslash B|\} \quad \forall A \subseteq E. \qquad \square$$

We note that the theory of sums of independence structures easily extends from two to k summands, say $\mathscr{E}_1 + \ldots + \mathscr{E}_k$; and the rank of $A(\subseteq E)$ in this structure (in the obvious notation) is equal to $\min_{B \subseteq A} \{\rho_1(B) + \ldots + \rho_k(B) + |A \backslash B|\}$. In particular, given an independence structure $\mathscr{E}$ on E with rank function ρ, its sum formed with itself k times is an independence structure with the rank of $A(\subseteq E)$ equal to $\min_{B \subseteq A} \{k\rho(B) + |A \backslash B|\}$.

This powerful result that the sum of a number of independence structures is again an independence structure has been particularly fruitful in affording simple proofs of results in vector space theory and graph theory, which had previously been quite difficult to establish. In this connection, we deduce from Theorem 2.17, for later use, the following corollary. It will be illustrated in an exercise at the end of the chapter and again in Chapter 3.

Corollary 2.18. *Let $(E, \mathscr{E})$ be an independence space with rank function ρ, and let k be a given positive integer. Then*

(i) E can be partitioned into k independent sets if and only if $|A| \leqslant k\rho(A)$ for each $A \subseteq E$;

(ii) there exist k pairwise-disjoint bases in E if and only if $k\rho(E) \leqslant k\rho(A) + |E \backslash A|$ for each $A \subseteq E$.

Proof. Let $\mathscr{E}'$ be the independence structure $\mathscr{E} + \ldots + \mathscr{E}$ with k summands, and let ρ' be its rank function.

(i) Now E can be partitioned into k members of $\mathscr{E}$ if and only if $E \in \mathscr{E}'$; i.e. if and only if

$$|E| = \rho'(E) = \min_{A \subseteq E} \{k\rho(A) + |E \backslash A|\}.$$

This is so if and only if

$$|E| \leqslant k\rho(A) + |E \backslash A|, \quad \text{i.e.} \quad |A| \leqslant k\rho(A) \quad \forall A \subseteq E.$$

(ii) Evidently E contains k pairwise-disjoint bases of $\mathscr{E}$ if and only if the set E has rank at least (in fact, equal to) $k\rho(E)$ in $\mathscr{E}'$; i.e. if and only if

$$k\rho(E) \leqslant \rho'(E) = \min_{A \subseteq E} \{k\rho(A) + |E \backslash A|\};$$

and the required result is immediate. □

Exercises

2.1 Let $E = \{1, 2, 3, 4, 5, 6\}$, and let $\mathscr{E}$ be the collection of all subsets of E of cardinality three or less with the exception of $\{1, 2, 5\}$, $\{1, 4, 6\}$, $\{2, 3, 6\}$ and $\{3, 4, 5\}$. Check that $(E, \mathscr{E})$ is an independence space.

2.2 Find six points in the plane and label them from 1 to 6 in such a way that the bases of the independence space in Exercise 2.1 are precisely the sets of three non-collinear points. [This shows that the independence structure in Exercise 2.1 consists of the affinely independent subsets of some set of six points of the plane.]

2.3 Let E be a given set and $\mathscr{E}$ a non-empty hereditary collection of subsets of E with the further property that, for every $A \subseteq E$, any two maximal members of $\mathscr{E}$ in A have the same cardinality. Prove that $\mathscr{E}$ is an independence structure on E.

*2.4 Let $(E, \mathscr{E})$ be an independence space, and let $x \in E$, $A \subseteq E$. We shall say that x *depends on* A and write $x|A$ if either $x \in A$, or $x \notin A$ and $\{x\} \cup B \notin \mathscr{E}$ for some independent subset B of A.

(i) Prove that, if $x|A$ and D is any maximal independent subset of A, then $x|D$.

(ii) Prove the following properties D(1), D(2), D(3) of the relation $|$.

D(1) $y_k|\{y_1, \ldots, y_m\}_{\neq}$ for $1 \leqslant k \leqslant m$.

D(2) If $m \geqslant 1$ and $x|\{y_1, \ldots, y_m\}_{\neq}$ whereas $x{\not|}\{y_2, \ldots, y_m\}_{\neq}$,

then $y_1|\{x_1 \ y_2, \ldots, y_m\}_{\neq}$.

D(3) If $x|\{y_1, \ldots, y_m\}_{\neq}$ and $y_k|\{z_1, \ldots, z_n\}_{\neq}$ for $1 \leqslant k \leqslant m$,

then $x|\{z_1, \ldots, z_n\}_{\neq}$.

Here $x, y_1, \ldots, y_m, z_1, \ldots, z_n$ denote elements of E. [The properties D(1), D(2), D(3) may be taken as alternative axioms upon which to build the theory of abstract independence: we should define a subset $A \subseteq E$ to be 'independent' if $x{\not|}A \backslash \{x\}$ for any $x \in A$. They are, in fact, essentially van der Waerden's original axioms (see Section 1.1).]

2.5 In the notation of Exercise 2.4, prove that, if $x \notin A$, then $x|A$ if and only if there exists a circuit C such that $x \in C \subseteq \{x\} \cup A$.

2.6 Let $E = \{1, 2, \ldots, 8\}$, and let $\mathscr{E}$ be the collection of all subsets of cardinality five or less with the exception of $\{1, 2, 5, 6, 7\}$, $\{1, 4, 5, 6, 8\}$, $\{2, 3, 6, 7, 8\}$ and $\{3, 4, 5, 7, 8\}$. Check that $(E, \mathscr{E})$ is an independence space.

2.7 In the independence space of Exercise 2.6, find two distinct circuits C_1 and C_2 such that $C_1 \triangle C_2$ itself contains no circuit. [In the next chapter, we shall meet a class of spaces for which this situation cannot arise.]

2.8 Let $(E, \mathscr{E})$ be the independence space of Exercise 2.6. Take in turn (i) $E' = \{6, 7, 8\}$ ($\in \mathscr{E}$), (ii) $E' = \{1, 2, 5, 6, 7\}$. List the bases of $\mathscr{E}_{\otimes E \, E'}$, the contraction of $\mathscr{E}$ away from E'.

2.9 Let V be a (finite) vector space over a field F and U a fixed linear subspace of V. Define a relation $\sim$ on V by the rule that $v \sim v'$ if and only if $v - v' \in U$. Then $\sim$ is an equivalence relation on V; we use $[v]$ to denote the equivalence class of V containing v and V/U to denote the collection of equivalence classes. It is easy to check that, for $v, w \in V$ and $\lambda \in F$, the operations defined unambiguously by

$$[v] + [w] = [v + w],$$
$$\lambda[v] = [\lambda v]$$

make V/U into a vector space over F, known as a *quotient space*. This exercise will now show that the concept of a contraction in an independence space is motivated by the idea of a quotient space.

Let $\mathscr{E}$ be the collection of linearly independent subsets of V, and assume that $[v_1], \ldots, [v_r]$ are distinct members of V/U. Show that $\{[v_1], \ldots, [v_r]\}$ is linearly independent in V/U if and only if $\{v_1, \ldots, v_r\} \in \mathscr{E}_{\otimes V \setminus U}$.

2.10 Let B be a basis of the independence space $(E, \mathscr{E})$, and suppose that $y \in E \setminus B$. Prove that

    (i) $B \cup \{y\}$ contains a circuit C;

    (ii) $y \in C$;

    (iii) C is the only circuit in $B \cup \{y\}$;

    (iv) $(B \cup \{y\}) \setminus \{x\}$ is a basis of $\mathscr{E}$ if and only if $x \in C$.

[The collection of all circuits C related to B in this way is the 'fundamental set of circuits of B'.]

**2.11 Let B, B' be two bases of the independence space $(E, \mathscr{E})$. Given $x \in B$, prove that there exists $y \in B'$ such that $(B \setminus \{x\}) \cup \{y\}$ and $(B' \setminus \{y\}) \cup \{x\}$ are *both* bases of $(E, \mathscr{E})$. [Hint: use Exercise 2.10(iv).]

2.12 Any set which contains a basis of an independence space $(E, \mathscr{E})$ is called a *spanning set* of $(E, \mathscr{E})$. Prove that, if A and A' are spanning sets of $(E, \mathscr{E})$ with $|A| > |A'|$, then $A \setminus \{a\}$ is also a spanning set for some $a \in A \setminus A'$. Make use of this result to give a direct proof of I(2) for the dual structure $\mathscr{E}^*$ on E.

2.13 Let $(E, \mathscr{E})$ be an independence space with rank function ρ, and let $A \subseteq E$. Show that the span of A is the set $\{x \in E : \rho(A \cup \{x\}) = \rho(A)\}$, and that its rank is $\rho(A)$.

*2.14 In an independence space $(E, \mathscr{E})$ of rank $n \geqslant 1$, a flat of rank $n - 1$ is called a *hyperplane*. Prove that

    (i) a set $A \subset E$ is a flat if and only if it is an intersection of hyperplanes;

    (ii) H is a hyperplane if and only if $E \setminus H$ is a circuit of the dual space $(E, \mathscr{E}^*)$.

2.15 An independence space is called *connected* if for each $x, y \in E$

with $x \neq y$ there exists a circuit C with $x, y \in C$. Show that the dual of a connected space is connected. [Hint: use Exercises 2.10(iv) and 2.14(ii).]

2.16 Prove that, if $\mathscr{E}, \mathscr{E}^*$ are dual structures on E and C, C^* are circuits of $\mathscr{E}, \mathscr{E}^*$ respectively, then $|C \cap C^*| \neq 1$. [We shall note later a stronger property which holds in the spaces considered in Chapter 3.]

2.17 Let $\mathscr{E}_1$ and $\mathscr{E}_1^*$ be dual structures on E_1, and $\mathscr{E}_2$ and $\mathscr{E}_2^*$ dual structures on E_2, where $E_1 \cap E_2 = \phi$. Write $E = E_1 \cup E_2$ and let $(\mathscr{E}_1 \oplus \mathscr{E}_2)^*$ be the dual of $\mathscr{E}_1 \oplus \mathscr{E}_2$ on E. Prove that

$$(\mathscr{E}_1 \oplus \mathscr{E}_2)^* = \mathscr{E}_1^* \oplus \mathscr{E}_2^*.$$

*2.18 Let $(E, \mathscr{E})$ be an independence space and $E' \subseteq E$. Establish the following relation between restrictions, contractions and duality

$$\mathscr{E}_{\otimes E \backslash E'} = (\mathscr{E}^* | E \backslash E')^*,$$

where $(\mathscr{E}^* | E \backslash E')^*$ denotes the dual of $\mathscr{E}^* | E \backslash E'$ in $E \backslash E'$.

2.19 Confirm that a truncation of an independence structure is itself an independence structure.

2.20 Let $(E, \mathscr{E})$ be an independence space with two of its bases B, B' lying in unique circuits $C = B \cup \{e\}$, $C' = B' \cup \{e'\}$, where $\{e, e'\} \nsubseteq C \cap C'$. Prove that $\mathscr{E}$ is not a proper truncation of an independence structure.

2.21 Prove that every independence space is the restriction of a properly truncated space.

2.22 An independence space is called a *circuit space* if every one of its bases lies in a circuit.

(i) Show that a proper truncation of an independence space is a circuit space.

(ii) Verify that the independence space in Exercise 2.6 is a circuit space, and use the result of Exercise 2.20 to show that it is not a proper truncation of an independence space.

*2.23 Show that a (finite) vector space of dimension $n \geqslant 2$ (with its linearly independent sets) is a circuit space but, for $n > 2$ is not a proper truncation of an independence space.

2.24 Give an example to show that the restriction of a circuit space is not necessarily a circuit space.

2.25 Show that a contraction of a circuit space $(E, \mathscr{E})$ away from $E' \subseteq E$ is itself a circuit space.

*2.26 Let $(E, \mathscr{E})$ be an independence space of rank r, and define inductively as follows collections $\mathscr{C}_0$, $\mathscr{C}_1$, ... and $\mathscr{C}$ of subsets of E.

I. $\mathscr{C}_0$ consists of the circuits of $\mathscr{E}$ of cardinality not exceeding r.

II. Given $\mathscr{C}_i$, let $\mathscr{C}_{i+1}$ consist of the members of $\mathscr{C}_i$ together with the circuits of $\mathscr{E}$ of cardinality $r+1$ which can be expressed in the form $(C \cup C') \backslash \{x\}$ for some $C, C' \in \mathscr{C}_i$ and $x \in C \cap C'$. [We show below in (i) that for some N, $\mathscr{C}_0 \subset \mathscr{C}_1 \subset \ldots \subset \mathscr{C}_N = \mathscr{C}_{N+1} = \ldots$.]

III. $\mathscr{C}$ consists of $\mathscr{C}_N$ together with all sets of $r+2$ elements of E which do not contain a member of $\mathscr{C}_N$.

Prove that

(i) an N exists as stated in II;

(ii) $\mathscr{C}$ is the collection of circuits of an independence structure, $\mathscr{E}'$ say, of rank r or $r+1$ on E;

(iii) $\mathscr{E}$ is a truncation of $\mathscr{E}'$;

(iv) $\mathscr{E}$ is a proper truncation if and only if $\mathscr{E} \neq \mathscr{E}'$.

[$\mathscr{E}'$ is called an *erection* of $\mathscr{E}$ if it properly truncates to $\mathscr{E}$. Here, then, we have a procedure for the construction of an erection of an independence structure.]

2.27 Consider the independence space of Exercise 2.1 and assign a weight $w(x)$ to each $x \in E$ by the rule

$$w(x) = 8x - x^2.$$

Write down the heaviest bases of $\mathscr{E}$ and confirm that they form the bases of an independence structure on E. Show also that the algorithm of Theorem 2.11 can lead to each of the heaviest bases.

2.28 Let $(E, \mathscr{E})$ be an independence space of rank r, and w a weight function defined on E. Prove that *every* heaviest basis of $\mathscr{E}$ can be obtained by the algorithm of Theorem 2.11. Prove, further, that if $B = \{x_1, \ldots, x_r\}_{\neq}$ and $B' = \{y_1, \ldots, y_r\}_{\neq}$ are heaviest bases with $w(x_1) \geqslant \ldots \geqslant w(x_r)$ and $w(y_1) \geqslant \ldots \geqslant w(y_r)$, then $w(x_i) = w(y_i)$ for $1 \leqslant i \leqslant r$.

2.29 Let E be a set and k be a non-negative integer. Show that the functions $\mu_1, \mu_2 : \mathscr{P}(E) \to \{0, 1, 2, \ldots\}$ defined by the rules

$$\mu_1(A) = \begin{cases} 0 & \text{if } A = \phi \\ k & \text{if } A \neq \phi \end{cases}, \quad \mu_2(A) = k|A| \quad \forall A \subseteq E$$

are submodular on E. In each case, find the independence structure induced on E.

2.30 Let $E = \{x_1, \ldots, x_m\}_{\neq}$ and $E' = \{y_1, \ldots, y_n\}_{\neq}$, where $m \leqslant n$, and let $\mathscr{E}'$ be an independence structure on E'. In each of the following cases, describe the structure $\mathscr{E}$ induced on E by the bipartite graph (E, Δ, E') (as in Theorem 2.16)

(i) $\Delta = \{x_1 y_1, \ldots, x_m y_m\}$;

(ii) $\Delta = \{x_i y_j : 1 \leqslant i \leqslant m, 1 \leqslant j \leqslant n\}$.

2.31 Let X be a subset of a vector space and let k be a positive integer. Under what conditions can X be partitioned into k linearly independent sets of vectors?

Decide which of the following subsets, X_1, X_2, of the real plane (i.e. the real vector space of pairs (x, y) of real numbers) can be partitioned into three linearly independent sets, and exhibit one such partition where possible

$$X_1 = \{(4, 0), (3, 1), (2, 2), (1, 3), (0, 4), (3, 3)\},$$
$$X_2 = \{(2, 0), (1, 1), (0, 2), (2, 2), (3, 3), (4, 4)\}.$$

2.32 Let $\mathscr{E}_1, \mathscr{E}_2$ be independence structures on a set E. Define their *product* $\mathscr{E}_1 \cdot \mathscr{E}_2$ by the rule

$$\mathscr{E}_1 \cdot \mathscr{E}_2 = (\mathscr{E}_1^* + \mathscr{E}_2^*)^*,$$

where $*$ denotes duality in E. Prove that

(i) $\mathscr{E}_1 \cdot \mathscr{E}_2$ is an independence structure on E;

(ii) $\mathscr{E}_1 \cdot \mathscr{E}_2 \subseteq \mathscr{E}_1 \cap \mathscr{E}_2 = \{A_1 \cap A_2 : A_1 \in \mathscr{E}_1, A_2 \in \mathscr{E}_2\}$.

Give a counter example to show that $\mathscr{E}_1 \cap \mathscr{E}_2$ is not necessarily an independence structure.

*__2.33__ Let E be a set and $\mathscr{C}, \mathscr{D}$ collections of non-empty subsets of E such that

(a) no proper subset of a member of $\mathscr{C}$ is a member of $\mathscr{C}$;

(b) no proper subset of a member of $\mathscr{D}$ is a member of $\mathscr{D}$;

(c) if $C \in \mathscr{C}$ and $D \in \mathscr{D}$, then $|C \cap D| \neq 1$;

(d) if $\{x\}, E_1, E_2$ form a partition of E, then either there exists $C \in \mathscr{C}$ with $x \in \mathscr{C} \subseteq \{x\} \cup E_1$ or there exists $D \in \mathscr{D}$ with $x \in D \subseteq \{x\} \cup E_2$.

Then $(E, \mathscr{C}, \mathscr{D})$ is called a *graphoid*.

(i) Show that, if $(E, \mathscr{E})$ is an independence space with $\mathscr{C}$ its collection of circuits and $\mathscr{D}$ the collection of circuits of its dual, then $(E, \mathscr{C}, \mathscr{D})$ is a graphoid.

(ii) Conversely, show that, if $(E, \mathscr{C}, \mathscr{D})$ is a graphoid, then $\mathscr{C}$ and $\mathscr{D}$ are the collections of circuits of dual independence structures on E.

CHAPTER THREE

Graphic spaces

3.1 The cycle and cutset structures of a graph

In this section we consider a graph $G = (V, E)$ and some independence structures on E arising naturally in the graph. We restrict attention for the present to a connected graph, for it is usually very easy to extend the concept to graphs which are not connected by considering each of their components in turn. The discussion of graphs which are not connected is left to the exercises at the end of the chapter.

Our first structure, $\mathscr{E}(G)$, on E is defined by the condition:

$$A \in \mathscr{E}(G) \Leftrightarrow A \text{ does not contain a cycle of } G.$$

$\mathscr{E}(G)$ is called the *cycle structure* of G (and $(E, \mathscr{E}(G))$ its *cycle space*).

Theorem 3.1 *The cycle structure $\mathscr{E}(G)$ of $G = (V, E)$ is an independence structure on E, and its circuits are precisely the cycles of G.*

Proof. We shall show that the collection $\mathscr{C}$ of cycles of G satisfies $C(1)$ and $C(2)'$, and therefore, by Theorem 2.6, that $\mathscr{C}$ is the collection of circuits of an independence structure on E, and that this structure is precisely $\mathscr{E}(G)$.

Now it is evident that no proper subset of a cycle is a cycle, and so $\mathscr{C}$ satisfies $C(1)$. Next, to establish $C(2)'$, let C, C' be different cycles of G with $C \cap C' \neq \phi$. Write $D = C \triangle C'$ and observe that the degree of a vertex in the graph (V, D) is equal to its degree in (V, C) plus its degree

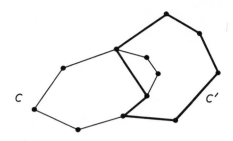

in (V, C') minus twice its degree in $(V, C \cap C')$; which is even. So if v is an endpoint of a member of D it is also the endpoint of another member of D. Now let $\{e_1, \ldots e_{n-1}\}_{\neq}$ be a maximal path in D with each e_i joining v_i to v_{i+1} say. Then, as v_n is the endpoint of some other member e of D, it follows (from the maximality assumption) that e $(\in D \setminus \{e_1, \ldots, e_{n-1}\})$ joins v_n to one of $v_i, \ldots, v_{n-1}$; to v_i say. Then $\{e_i, e_{i+1}, \ldots, e_{n-1}, e\}_{\neq}$ is a cycle of G contained in D : in particular, it is a member of $\mathscr{C}$ contained in $(C \cup C') \setminus \{x\}$ for any $x \in C \cap C'$. Hence $C(2)'$ is established for $\mathscr{C}$. (In fact, we shall see in Section 3.2 that $C \triangle C'$ is a disjoint union of cycles.) As stated above, the required result now follows from Theorem 2.6.   □

In our connected graph $G = (V, E)$, let $T \subseteq E$ be minimal with respect to the property that the graph (V, T) is connected. It follows, then, that T contains no cycle (for if it did, then the deletion of an edge of that cycle would still leave a connected graph), and hence that T is a tree. Conversely, (V, T) connected and T a tree ensures minimality. So a set of edges $T \subseteq E$, minimal such that (V, T) is connected, is precisely a set of edges with (V, T) connected and containing no cycle. Such a set T will be called a *spanning tree* of G. We shall see in the exercises that, for any spanning tree T of $G = (V, E)$, $|T| = |V| - 1$.

Example

We illustrate on the next page a graph G and one of its spanning trees.

Theorem 3.2. *The spanning trees of G form the bases of the cycle structure of G.*

Proof. As usual, let $G = (V, E)$. Assume first that $T \subseteq E$ is a basis of $\mathscr{E}(G)$. Then $T \in \mathscr{E}(G)$ and so contains no cycle of G. Also (V, T) is connected. For, otherwise, the addition of a member of E joining two

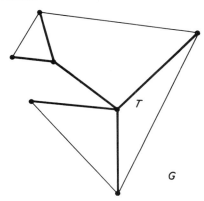

components of (V, T) would give a larger subset of E than T still containing no cycle, and hence still in $\mathscr{E}(G)$. Thus T is a spanning tree of G, as required. Recall also that $|T| = |V| - 1$; and hence that $\mathscr{E}(G)$ has rank $|V| - 1$.

Conversely, if T is a spanning tree of G, then it contains no cycles and so is a member of $\mathscr{E}(G)$. Also T has cardinality equal to the rank of $\mathscr{E}(G)$; and so T is a basis of $\mathscr{E}(G)$ as required.   □

Yet another characteristic property of spanning trees follows from this theorem: namely, a spanning tree is precisely a maximal set of edges containing no cycle.

We now turn to a seemingly unrelated structure, although the form of our development of it will very closely parallel that of the cycle structure. In our connected graph $G = (V, E)$, we say that $A \subseteq E$ *disconnects* G if the graph $(V, E \backslash A)$ is not connected. Further, a *cutset* of G is a minimal disconnecting set of edges. Evidently, in view of its minimality, a cutset disconnects G into just two components.

Example

In the graph shown overleaf, examples of cutsets are $\{1, 2\}$, $\{2, 3, 5, 6\}$ and $\{1, 3, 5, 7, 8\}$.

The structure $\mathscr{E}^*(G)$ on E which we now consider is defined by the condition:

$$A \in \mathscr{E}^*(G) \Leftrightarrow A \text{ contains no cutset of } G.$$

$\mathscr{E}^*(G)$ is called the *cutset structure* of G (and $(E, \mathscr{E}^*(G))$ its *cutset space*).

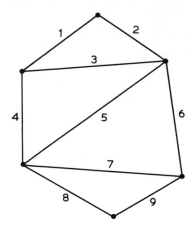

Theorem 3.3 (cf. Theorem 3.1.) *The cutset structure $\mathscr{E}^*(G)$ of $G = (V, E)$ is an independence structure on E, and its circuits are precisely the cutsets of G.*

Proof. We shall show that the collection $\mathscr{C}$ of cutsets of G satisfies $C(1)$ and $C(2)'$; and the required result will then follow at once from Theorem 2.6. By the minimality of the disconnecting sets in $\mathscr{C}$, it is clear that no proper subset of a member of $\mathscr{C}$ is itself a member of $\mathscr{C}$; and hence $\mathscr{C}$ satisfies $C(1)$. Now let C, C' be different members of $\mathscr{C}$, and assume that the components of $(V, E\backslash C)$ have vertex sets V_1, V_2 and that those of $(V, E\backslash C')$ have vertex sets V_3, V_4. Then the whole of G is shown in the following diagrammatic representation:

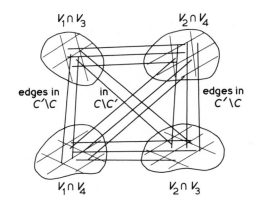

Now the edges from the top of the figure to the botton form the (non-empty) set $D = C \triangle C'$ ($= (C' \backslash C) \cup (C \backslash C')$), and so certainly D discon-nects G. Hence D contains a cutset of G which, in particular, is a member of $\mathscr{C}$ in $(C \cup C') \backslash \{x\}$ for any $x \in C \cap C'$. Therefore $C(2)'$ is established for $\mathscr{C}$. (In fact, we shall see in Section 3.2 that $C \triangle C'$ is a disjoint union of cutsets.) As stated above, the required result now follows from Theorem 2.6.   □

Theorem 3.4. (cf. Theorem 3.2.) *The complements in E of the spanning trees of* $G = (V, E)$ *form the bases of the cutset structure of G.*

Proof. We simply note that S is a basis of $\mathscr{E}^*(G)$ if and only if it is a maximal set containing no cutset; equivalently, if it is maximal with respect to the property that $(V, E \backslash S)$ is connected. This is, in turn, equivalent to saying that $E \backslash S$ is minimal with respect to the property that $(V, E \backslash S)$ is connected, i.e. that $E \backslash S$ is a spanning tree of G.   □

Of course, it is an immediate consequence of Theorem 3.2 and of our work on duality in Chapter 2 that the complements of the spanning trees of G form the bases of an independence space; and we could have taken this as our starting point for a discussion of $\mathscr{E}^*(G)$. The following corollary is now immediate.

Corollary 3.5. *The cycle and cutset structures of a graph are dual independence structures.*   □

It is for this reason that we have adopted the * notation for the cutset structure. We shall speak of $(E, \mathscr{E}(G))$ and $(E, \mathscr{E}^*(G))$ as the *graphic spaces* associated with G.

Example

For the graph G illustrated we list the cycles and cutsets, and the bases of $\mathscr{E}(G)$ and $\mathscr{E}^*(G)$.

Cycles	Cutsets
$\{1, 2, 3\}, \{5, 6\}$	$\{1, 2\}, \{1, 3\}, \{2, 3\}, \{4\},$
	$\{5, 6\}, \{7\}.$

Bases of $\mathscr{E}(G)$

$\{1, 2, 4, 5, 7\}, \{1, 2, 4, 6, 7\},$
$\{1, 3, 4, 5, 7\}, \{1, 3, 4, 6, 7\},$
$\{2, 3, 4, 5, 7\}, \{2, 3, 4, 6, 7\}.$

Bases of $\mathscr{E}^*(G)$

$\{3, 6\}, \{3, 5\},$
$\{2, 6\}, \{2, 5\},$
$\{1, 6\}, \{1, 5\}.$

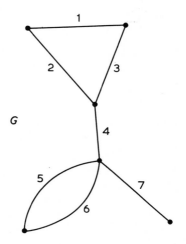

We shall now investigate another link between the cycle and cutset structures in the case of certain graphs which can be drawn in the plane in a rather special way. A graph is *planar* if it can be drawn in the plane in such a way that no two edges meet except at a common endpoint of them both; and such a 'representation' of a graph is called a *planar representation*. So, for example, the complete graph K_4 and the complete bipartite graph $K_{2,3}$ are planar as the following figures convincingly show:

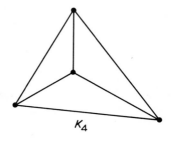

K_4

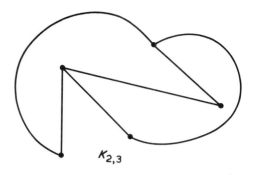

$K_{2,3}$

On the other hand, we shall see later that K_5 and $K_{3,3}$ are not planar.

Now, given a planar graph $G = (V, E)$ and a planar representation of it, in a natural way the representation divides the plane into separate regions or *faces*. So, for example, the above planar representations of K_4 and $K_{2,3}$ have four and three faces, respectively. Note also that, given any edge in the planar representation, there is a face on each side of it (and, in some cases, these two faces coincide). From a planar representation of G we can now define a new (planar) graph G^*, called a *geometric dual* of G. First, we take as the vertices of G^* one point in each of the faces of the representation of G. Next, each edge e of G has one face on each side of it and corresponding to this edge we create an edge e^* of G^* by joining those vertices of G^* which lie in the faces on opposite sides of e (and e^* may be a loop). This seemingly complicated process becomes transparent with the aid of the example illustrated overleaf.

We note that any geometric dual of our planar graph G is connected. This is because, in any partitioning of the plane into faces, any face can be reached from any other face via pairwise-adjacent faces. We stress that a dual depends on the particular representation of G, and hence a planar graph can have several geometric duals. It turns out that there is a very close connection between the cycles of a planar graph and the cutsets of any one of its geometric duals (and vice versa). We shall now examine this connection and hence deduce the further link between the cycle and cutset structures promised above. In the proof, we shall

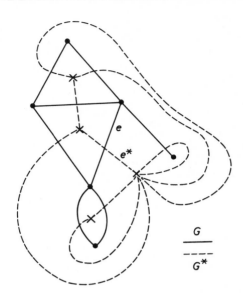

use the notion of an *arc* in the plane. For our purposes this is simply the image of a continuous injection f of the closed interval $[0, 1]$ into the plane; a continuous route from $f(0)$ to $f(1)$ which does not cross itself.

Theorem 3.6. *Let $G = (V, E)$ be a planar graph with a geometric dual G^* $= (V^*, E^*)$ in which each $e \in E$ is associated with an $e^* \in E^*$ in the natural way described above (from a given planar representation of G). Then C is a cycle of G if and only if $C^*(= \{e^* : e \in C\})$ is a cutset of G^*.*

Proof. We shall show that $D \subseteq E$ contains a cycle of G if and only if D^* disconnects G^*. From this it is evident that C is a cycle of G if and only if C^* disconnects G^* but no proper subset of C^* has this property; i.e. if and only if C^* is a cutset of G^*.

So let us assume first that $D \subseteq E$ contains a cycle C of G. Consider the natural planar representation of G^* superimposed upon that of G from which it arises (as illustrated in the previous figure), and let V_1^* be the (non-empty) set of those members of V^* inside C, and $V_2^* = V^* \backslash V_1^*$ be the (non-empty) set of those outside C. Then clearly $e^* \in C^*$ if and only if e^* crosses the cycle C; i.e. if and only if e^* joins a vertex of V_1^* to a vertex of V_2^*. It follows that C^* (and hence D^*) disconnects G^*.

Conversely, assume that D does not contain a cycle of G. Then, again in the naturally associated representations of G and G^*, between any two distinct vertices v^*, w^* of G^* there is an arc in the plane which avoids the edges in D and all the vertices of G. Assume that this arc meets, in order, the following faces of G (specified by the vertices v_i^* of G^* which they contain) and edges of G, as listed:

$$v^*, e_1, v_1^*, e_2, v_2^*, e_3, \ldots, e_{n-1}, v_{n-1}^*, e_n, w^*.$$

Clearly, the fact that the arc passes from face v_i^* to v_{i+1}^* crossing edge e_{i+1} means that v_i^* and v_{i+1}^* are joined by e_{i+1}^* ($\in E^* \backslash D^*$), as in the

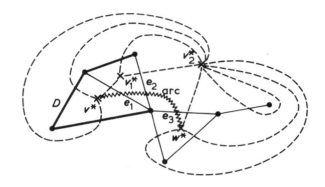

figure. Therefore

$$v^*, e_1^*, v_1^*, e_2^*, v_2^*, e_3^*, \ldots, e_{n-1}^*, v_{n-1}^*, e_n^*, w^*$$

is a sequence of vertices and edges of $(V^*, E^* \backslash D^*)$ starting at v^* ending at w^* with each e_i^* joining its two neighbouring vertices in the sequence. A minimal such sequence is a path in $(V^*, E^* \backslash D^*)$ joining v^* to w^*; which shows that D^* does not disconnect G^*. $\square$

We can now see (as the reader has probably already been suspecting) that the cycle structure of G is simply a copy on E of the cutset structure of G^* on E^*; i.e. there is an 'isomorphism' between these two structures.

Corollary 3.7. *Let $G = (V, E)$ be a planar graph with a geometric dual*

$G^* = (V^*, E^*)$ in which each $e \in E$ is associated with an $e^* \in E^*$ in the way described.⁻ Then $A \in \mathscr{E}(G)$ if and only if $A^*(= \{e^* : e \in A\}) \in \mathscr{E}^*(G^*)$.

Proof. We simply note that $A \in \mathscr{E}(G)$ if and only if A contains no cycle of G which, by Theorem 3.6, happens if and only if A^* contains no cutset of G^*; i.e. if and only if $A^* \in \mathscr{E}^*(G^*)$. □

Since $\mathscr{E}(G)$ and $\mathscr{E}^*(G^*)$ are isomorphic structures, it follows at once, of course, that so are their duals $\mathscr{E}^*(G)$ and $\mathscr{E}(G^*)$.

Corollary 3.8. If G^* and $G^{*\prime}$ are both geometric duals of the planar graph G, then $\mathscr{E}(G^*)$ and $\mathscr{E}(G^{*\prime})$ are isomorphic structures.

Proof. Each is a copy of $\mathscr{E}^*(G)$. □

We would emphasize to the reader that our discussion of planar graphs and their geometric duals has led to the striking result that the cycle structure of a planar graph may be interpreted as the cutset structure of another planar graph (and, likewise, its cutset structure may be interpreted as a cycle structure). In fact, the converse of this result is also true; so that planar graphs may be characterized by this property. The proof of this converse assertion is, however, much more difficult and is beyond our present scope.

3.2 Connections with vector spaces

When we speak here of a vector space, we shall understand that its collection of *linearly* independent sets is being considered. One of our objectives in this section is to show how the graphic spaces $(E, \mathscr{E}(G))$ and $(E, \mathscr{E}^*(G))$ are essentially copies of subsets of a vector space over the field $GF(2)$. A somewhat similar discussion appears in Chapter 4 for the 'transversal spaces' which we consider there. The general subject of representing an independence space as part of a vector space will be looked at briefly in Chapter 5.

Recall that $GF(2)$ is the field $\{0, 1\}$ with addition and multiplication defined by the tables

+	0	1
0	0	1
1	1	0

.	0	1
0	0	0
1	0	1

For a fixed positive integer n, the row (or column) matrices of length n with elements 0, 1 form a vector space $V_n = (GF(2))^n$ over $GF(2)$.

Now let $G = (V, E)$ be a graph. We shall suppose, as in Section 3.1, that G is connected; and we assume also that it is not the trivial graph in which $E = \phi$. Let us write $E = \{e_1, \ldots, e_n\}_\neq$, and let us suppose that the members of E are ordered once and for all. We set up a bijective (i.e. one-to-one) correspondence (written $\leftrightarrow$) between subsets A of E and vectors of V_n by the rule:

$$A \subseteq E \leftrightarrow (\alpha_1, \ldots, \alpha_n), \quad \text{where} \quad \alpha_i = \begin{cases} 1 & \text{if } e_i \in A \\ 0 & \text{if } e_i \notin A. \end{cases}$$

In particular, if we identify $\{e_i\}$ with e_i, then

$$e_i \leftrightarrow (0, \ldots, 0, \overset{(i)}{1}, 0, \ldots, 0).$$

We begin our investigations with a study of two particular subspaces X and Y of V_n spanned (or generated) respectively by the vectors corresponding to the cycles and the cutsets of G.

Theorem 3.9. *A member of V_n is in X if and only if it corresponds to a disjoint union (possibly empty) of cycles of G.*

Proof. We first prove, by induction on $|D|$, that if $D(\subseteq E)$ is such that, in the graph (V, D), each vertex has even degree, then D is a disjoint union of cycles. The result is trivial if $|D| = 0$. Assume, then, that $|D| > 0$, that D has the given property, and that the result is true for all subsets of E smaller than D. Then each vertex of G which is the endpoint of a member of D is the endpoint of at least two members of D (or possibly of a loop in D), and it follows (as in the proof of Theorem 3.1) that D contains a cycle, say C. But then, in $(V, D\backslash C)$, each vertex still has even degree and so, by the induction hypothesis, $D\backslash C$ (and hence D) is a disjoint union of cycles. The converse result is also true; for if D is a disjoint union of cycles and $v \in V$ is used in exactly r of these cycles, then its degree in (V, D) is $2r$. Hence D is a disjoint union of cycles if and only if, in (V, D), each vertex has even degree.

Now let X' be the set of vectors of V_n which correspond to subsets D of E with the property that each vertex in (V, D) has even degree. We want to show that $X' = X$. Certainly the zero vector belongs to X', so $X' \neq \phi$. Next, let $x, x' \in X'$ and (say) $D \leftrightarrow x$, $D' \leftrightarrow x'$. Then, from the addition rule in $GF(2)$, $D \triangle D' \leftrightarrow x + x'$. But the degree of any vertex in

$(V, D \triangle D')$ is equal to its degree in (V, D) plus its degree in (V, D') minus twice its degree in $(V, D \cap D')$, which is even. Hence $x + x' \in X'$, and it follows that X' is a subspace of V_n. From this fact, since every vector corresponding to a cycle is clearly in X', it follows that $X \subseteq X'$. Finally, if $x \in X'$, then x corresponds to a disjoint union of cycles $C_1, \ldots, C_r$ (say) and, if $C_1 \leftrightarrow x_1, \ldots, C_r \leftrightarrow x_r$, with $x_1, \ldots, x_r \in X$, then $x = x_1 + \ldots + x_r \in X$; and therefore $X' \subseteq X$. □

The exactly analogous situation holds for Y, as stated below.

Theorem 3.10. *A member of V_n is in Y if and only if it corresponds to a disjoint union (possibly empty) of cutsets of G.*

Proof. If D is a cutset of G and the components of $(V, E \backslash D)$ have vertex sets V_1 and V_2, then each cycle of G, being a closed path, must contain an even number of edges joining V_1 and V_2, so that $|D \cap C|$ is even. The corresponding result when D is a disjoint union of cutsets is immediate. Conversely, assume that D has the property that $|D \cap C|$ is even for every cycle C of G. We prove, by induction on $|D|$, that D is a disjoint union of cutsets. The result is trivial if $|D| = 0$. Suppose, then, that $|D| > 0$, that D has the given property, and that the result is true for all subsets of E smaller than D. Now $(V, E \backslash D)$ must be disconnected; for, if not, $E \backslash D$ would contain a spanning tree T of (V, E) and, for $d \in D$, $T \cup \{d\}$ would contain a cycle C with $|C \cap D| = |\{d\}| = 1$. Therefore D contains a cutset D' and, since D' meets every cycle in an even number of edges, so does $D \backslash D'$. So, by the induction hypothesis, $D \backslash D'$ (and hence D) is a disjoint union of cutsets.

Now let Y' be the set of vectors of V_n which correspond to subsets D of E with the property that they meet each cycle of G in an even number of edges. We want to show that $Y' = Y$. Certainly the zero vector belongs to Y', so $Y' \neq \phi$. Next, let $y, y' \in Y$ and (say) $D \leftrightarrow y$, $D' \leftrightarrow y'$. Then $D \triangle D' \leftrightarrow y + y'$; and, if C is a cycle of G,

$$|C \cap (D \triangle D')| = |C \cap D| + |C \cap D'| - 2|C \cap (D \cap D')|,$$

which is even. Hence $y + y' \in Y'$ and it follows that Y' is a subspace of V_n. From this fact, since every vector corresponding to a cutset is clearly in Y', we deduce that $Y \subseteq Y'$. Finally, if $y \in Y'$, then y corresponds to a disjoint union of cutsets $D_1, \ldots, D_r$ (say) and, if $D_1 \leftrightarrow y_1, \ldots, D_r \leftrightarrow y_r$, with $y_1, \ldots, y_r \in Y$, then $y = y_1 + \ldots + y_r \in Y$; and therefore $Y' \subseteq Y$. □

There is another natural spanning set for Y which it is useful for us to note. For each $v \in V$, let us define the *edge-set through v* to be the set of those edges of G which have precisely one endpoint at v. (So loops at v do not belong to the edge-set through v.)

Theorem 3.11. *The vectors corresponding to the edge-sets through the vertices of G span Y.*

Proof. Let $v \in V$ have edge-set $F \subseteq E$. Then $(V, E \backslash F)$ has components on vertex sets $\{v\}$, $V_1, \ldots, V_r$ (say). Clearly the edges joining v to any one of the V_i form a cutset of G, and F is the disjoint union of these r cutsets. Hence a vector corresponding to an edge-set through a vertex is a member of Y. Now let D be a cutset of G, and suppose that the vertex sets of the two components into which it disconnects G are V_1 and V_2. Then the edges of D are precisely those with one endpoint in V_1 and the other in V_2; otherwise a proper subset of D would disconnect G. Let the (distinct) vectors corresponding to the edge-sets through the vertices of V_1 be $y_1, \ldots, y_s$ and let the vector corresponding to D be y. Now the vector $y_1 + \ldots + y_s$ evidently has a '1' in the ith position if and only if e_i is in odd number, and so in precisely *one*, of the edge-sets through vertices of V_1, i.e. if and only if $e_i \in D$; and so $y = y_1 + \ldots + y_s$. From this we readily deduce that any member of Y can be expressed as a sum of vectors corresponding to edge-sets through vertices. $\qquad\square$

For $x = (\alpha_1, \ldots, \alpha_n)$, $y = (\beta_1, \ldots, \beta_n) \in V_n$, we recall from Section 1.3 that $x \cdot y = \sum_{i=1}^{n} \alpha_i \beta_i$, where now this calculation is carried out in $GF(2)$ (so that, for example, $(0, 1, 1, 1, 1) \cdot (0, 1, 0, 1, 1) = 1$). It is at once clear that $x \cdot y = 0$ if and only if there is an even number of '1s' in the same corresponding positions in x and y. Since a cycle and a cutset have an even number of edges in common, the following theorem will not prove surprising.

Theorem 3.12. *With X and Y as above,*

$$Y = \{y \in V_n : y \cdot x = 0 \quad for\ all \quad x \in X\}$$

and

$$X = \{x \in V_n : x \cdot y = 0 \quad for\ all \quad y \in Y\}.$$

Proof. Let $x_1, \ldots, x_t$ be precisely those vectors corresponding to all the cycles $C_1, \ldots, C_t$ of G. Then, exactly as in the proof of Theorem 3.10, D corresponds to a member of Y if and only if each of the numbers $|D \cap C_1|, \ldots, |D \cap C_t|$ is even. It follows that $y \in Y$ if and only if each of the products $y \cdot x_1, \ldots, y \cdot x_t$ is zero. Since $x_1, \ldots, x_t$ span X, we conclude that $y \in Y$ if and only if $y \cdot x = 0$ for each $x \in X$.

The second assertion is now an immediate consequence of Corollary 1.4. Alternatively, we may give a graph-theoretic argument. For, let $y_1, \ldots, y_s$ be precisely those vectors corresponding to the edge-sets $F_1, \ldots, F_s$ through all the vertices of G. Then, for $D \subseteq E$, the degree in (V, D) of a vertex v with edge-set F_i is equal to $|D \cap F_i|$ plus twice the number of loops at v in D. Hence, as in the proof of Theorem 3.9, $x \in X$ if and only if x corresponds to $D (\subseteq E)$, where each of $|D \cap F_1|, \ldots, |D \cap F_s|$ is even; i.e. if and only if each of the products $x \cdot y_1, \ldots, x \cdot y_s$ is zero. By Theorem 3.11, $y_1, \ldots, y_s$ span Y, and so $x \in X$ if and only if $x \cdot y = 0$ for each $y \in Y$. $\square$

We now come to our promised vector representations of $\mathscr{E}(G)$ and $\mathscr{E}^*(G)$.

Theorem 3.13. *Let X, Y be as above, and let $\{y_1, \ldots, y_s\}_{\neq}$ be any spanning set for Y. Let $J(G)$ be the $s \times n$ matrix whose rows are the vectors $y_1, \ldots, y_s$, and set up a correspondence between its columns $1, \ldots, n$ and the edges $e_1, \ldots, e_n$ of G (or, equivalently, 'label' the columns $e_1, \ldots, e_n$). Then $A \in \mathscr{E}(G)$ if and only if the columns of $J(G)$ labelled by the members of A are distinct and form a linearly independent set (over $GF(2)$).*

Proof. Let $A = \{e_1, \ldots, e_r\}_{\neq}$ for ease of labelling. Then A corresponds to the vector $(\underbrace{1, \ldots, 1}_{r}, 0, \ldots, 0)$ of V_n, and

$A \in \mathscr{E}(G) \Leftrightarrow A$ contains no cycle of G

$\Leftrightarrow A$ contains no non-empty set corresponding to a vector in X (by Theorem 3.9)

$\Leftrightarrow$ the condition that $(\lambda_1, \ldots, \lambda_r, 0, \ldots, 0) \in X$ implies that $\lambda_1 = \ldots = \lambda_r = 0$

$\Leftrightarrow$ the condition that $z = (\lambda_1, \ldots, \lambda_r, 0, \ldots, 0)$ satisfies $z \cdot y_1 = \ldots = z \cdot y_s = 0$ implies that $\lambda_1 = \ldots = \lambda_r = 0$ (by Theorem 3.12)

$$\Leftrightarrow \text{the condition that } J(G) \begin{pmatrix} \lambda_1 \\ \vdots \\ \lambda_r \\ 0 \\ \vdots \\ 0 \end{pmatrix} = \begin{pmatrix} 0 \\ \vdots \\ 0 \end{pmatrix}$$

implies that $\lambda_1 = \ldots = \lambda_r = 0$

$\Leftrightarrow$ the first r columns of $J(G)$ are distinct and form a linearly independent set

$\Leftrightarrow$ the columns of $J(G)$ corresponding to members of A are distinct and form a linearly independent set. $\square$

The most natural choice of $J(G)$ in Theorem 3.13 is to take for its rows those vectors of V_n which correspond to the edge-sets $F_1, \ldots, F_s$ (say) through the vertices of G. In the situation when G has no loops, such a matrix, known as an *incidence matrix* of G, is easy to characterize: let $V = \{v_1, \ldots, v_s\}_{\neq}$ and take $v_1, \ldots, v_s$ in this order; then the (i, j)th element of the corresponding incidence matrix of G is 1 if v_i is an endpoint of e_j, and 0 otherwise.

Example

An incidence matrix of the graph

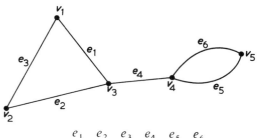

is

$$\begin{array}{c@{\quad}ccccccc} & e_1 & e_2 & e_3 & e_4 & e_5 & e_6 \\ v_1 & \left(1\right. & 0 & 1 & 0 & 0 & 0 \\ v_2 & 0 & 1 & 1 & 0 & 0 & 0 \\ v_3 & 1 & 1 & 0 & 1 & 0 & 0 \\ v_4 & 0 & 0 & 0 & 1 & 1 & 1 \\ v_5 & 0 & 0 & 0 & 0 & 1 & \left.1\right) \end{array}$$

Corollary 3.14. *If G has no loops, then $A \in \mathscr{E}(G)$ if and only if the corresponding columns of an incidence matrix of G are distinct and form a linearly independent set (over $GF(2)$).*   □

Of course, either Theorem 3.13 or Corollary 3.14 provides an alternative proof that $\mathscr{E}(G)$ is an independence structure. We have a similar situation for $\mathscr{E}^*(G)$.

Theorem 3.15. *Let X, Y be as above, and let $\{x_1, \ldots, x_t\}_{\neq}$ be any spanning set for X. Let $K(G)$ be the $t \times n$ matrix whose rows are the vectors $x_1, \ldots, x_t$, and set up a correspondence between its columns $1, \ldots, n$ and the edges $e_1, \ldots, e_n$ of G (or, equivalently 'label' the columns $e_1, \ldots, e_n$). Then $A \in \mathscr{E}^*(G)$ if and only if the columns of $K(G)$ labelled by the members of A are distinct and form a linearly independent set (over $GF(2)$).*

Proof. This follows the proof of Theorem 3.13 so closely that we omit the details.   □

We remark that there is no such natural choice of $K(G)$ corresponding to the choice of an incidence matrix for $J(G)$.

3.3 Applications of independence theory to graphs

We shall now exhibit some purely graph-theoretic results in the light of the theory of independence. Only a few samples will be given of what can be achieved. Our first illustration is particularly simple, and the result is the well-known formula of Euler connecting the numbers of vertices, edges and faces of a graph drawn in the plane.

Theorem 3.16. *Let $G = (V, E)$ be a connected planar graph, and F the collection of faces of any particular planar representation of G. Then*

$$|F| = |E| - |V| + 2.$$

Proof. Let $G^* = (V^*, E^*)$ be the geometric dual of G corresponding to the planar representation of G considered. Then, in particular, $|V^*| = |F|$. Further, $\mathscr{E}(G^*)$ and $\mathscr{E}^*(G)$ are isomorphic structures and it follows that their ranks are equal. Therefore

$$|V^*| - 1 = |E| - |V| + 1,$$

and hence

$$|F| = |E| - |V| + 2.$$ □

We have remarked earlier that K_5 and $K_{3,3}$ are not planar graphs. We give a proof of this for $K_{3,3}$ in our next theorem. (A proof of the corresponding result for K_5 is left as an exercise at the end of the chapter.)

Theorem 3.17. $K_{3,3}$ *is not planar.*

Proof. Let $K_{3,3}$ be the graph $G = (V, E)$ and assume that it is planar. Then it has a geometric dual $G^* = (V^*, E^*)$ whose cycles correspond in the usual way to the cutsets of G. Also, in this case, $|E^*| = |E| = 9$ and (as in the proof of Theorem 3.16) $|V^*| = |E| - |V| + 2 = 5$. Next, G^* has no loops or repeated edges (for any of these would produce in G^* a cycle of fewer than three edges, whereas G has no such cutset). Hence G^* has five vertices and nine edges and is precisely the complete graph on five vertices with one edge removed. In particular, G^* has two vertices of degree 3, and so E^* is not a disjoint union of cycles of G^*. However, E *is* a disjoint union of cutsets of G (namely the three cutsets of three edges each, as shown in the figure). This contradiction completes the proof that $K_{3,3}$ cannot be planar.

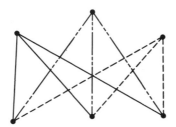

□

In Chapter 2, Theorem 2.11, we obtained an algorithm for constructing a heaviest basis of an independence space. Its interpretation for cycle spaces gives the following result.

Theorem 3.18. *Let $G = (V, E)$ be a connected graph, and associate with each edge of G a non-negative real number (called the 'weight' of this edge). Then a spanning tree of G of maximum total weight may be constructed by the following procedure:*
   I. *Put $T_0 = \phi$.*

II. *Suppose that* $0 \leqslant i < |V| - 1$ *and that the sets* $T_0, \ldots, T_i$ *have been constructed. Let* T_{i+1} *be formed by adjoining to* T_i *an element* $e \in E \backslash T_i$ *with* e *of largest possible weight subject to the condition that* $T_i \cup \{e\}$ *contains no cycle.*
Then $T_{|V|-1}$ *is a spanning tree of* G *of maximum total weight.*

Proof. This construction is precisely that of Theorem 2.11 in the case $\mathscr{E} = \mathscr{E}(G)$ (and $\rho(E) = |V| - 1$). Hence the process will lead to a heaviest basis of $\mathscr{E}(G)$, i.e. a spanning tree of G of maximum total weight. □

Independence theory has illuminated many graph-theoretic situations and the connections between them. Our examples so far have been particularly simple and have done scant justice to the power of independence-theoretic methods. By contrast, the application to which we now turn in conclusion is quite a deep one. Indeed, in Theorem 3.19 we see an example of one of the striking achievements of independence theory, namely to provide simple proofs of certain so-called 'covering and packing' results for graphs which previously had only been established with considerable effort and ingenuity. They appear now as immediate consequences of the work in Section 2.4 on sums of independence structures. It is necessary to introduce one further item of notation: for a subset A of the edge-set of our connected graph $G = (V, E)$, let $c(A)$ denote the number of components of (V, A). As we shall see in Exercise 3.3, the rank function ρ of the cycle structure of G is given by the formula

$$\rho(A) = |V| - c(A) \quad \forall A \subseteq E.$$

Theorem 3.19. *Let* $G = (V, E)$ *be a connected graph, and let* k *be a given positive integer. Then*
(i) *E can be partitioned into* k *forests if and only if, for each* $A \subseteq E$,

$$|A| \leqslant k(|V| - c(A));$$

(ii) *G possesses* k *pairwise edge-disjoint spanning trees if and only if, for each* $A \subseteq E$,

$$k(c(A) - 1) \leqslant |E \backslash A|.$$

Proof. By Corollary 2.18 applied to $\mathscr{E} = \mathscr{E}(G)$, with rank function ρ, (i) E can be partitioned into k members of $\mathscr{E}(G)$ (i.e. forests of G) if and only if, for each $A \subseteq E$, $|A| \leqslant k\rho(A) = k(|V| - c(A))$; and (ii) E possesses k pairwise-disjoint bases of $\mathscr{E}(G)$ (i.e. spanning trees of G) if and only if, for each $A \subseteq E$,

$$k\rho(E) = k(|V| - c(E)) \leqslant k\rho(A) + |E \backslash A|$$
$$= k(|V| - c(A)) + |E \backslash A|,$$

or equivalently

$$k(c(A) - 1) \leqslant |E \backslash A|. \qquad \square$$

Exercises

3.1 Prove that, if T is a spanning tree of the connected graph $G = (V, E)$, then $|T| = |V| - 1$.

3.2 Let $G = (V, E)$ be a connected graph and T a spanning tree of G. Show that
   (i) for each $e \in E \backslash T$, there is a unique cycle in $T \cup \{e\}$;
   (ii) for each $e \in T$, there is a unique cutset in $(E \backslash T) \cup \{e\}$.

3.3 Find the rank functions of the cycle and cutset structures of a connected graph $G = (V, E)$.

3.4 Show that K_4 is a geometric dual of itself. Find other graphs with this property.

3.5 Prove that K_5, the complete graph on five vertices, is not planar.

3.6 The figures below illustrate two different planar representations of the same graph G. Obtain the associated geometric duals, and show that they are essentially different graphs, say G^*, $G^{*\prime}$. Verify directly that, on the other hand, $\mathscr{E}(G^*)$ and $\mathscr{E}(G^{*\prime})$ are isomorphic.

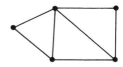 

3.7 $G = (V, E)$ be a graph, not necessarily connected, and define the collection $\mathscr{E}(G)$ of subsets of E by the condition:

$$A \in \mathscr{E}(G) \Leftrightarrow A \text{ contains no cycle of } G.$$

Show that $\mathscr{E}(G)$ is the direct sum of the cycle structures on the components of G. Hence, find the rank of E in $\mathscr{E}(G)$.

3.8 Let $G = (V, E)$ be a graph, not necessarily connected, and define the collection $\mathscr{E}^*(G)$ of subsets of E by the condition:

> $A \in \mathscr{E}^*(G) \Leftrightarrow$ the number of components of G is equal to the number of components of $(V, E \backslash A)$.

Show that $\mathscr{E}^*(G)$ is the direct sum of the cutset structures on the components of G. Hence, find the rank of E in $\mathscr{E}^*(G)$.

3.9 Prove that, for a graph $G = (V, E)$ which is not necessarily connected, $\mathscr{E}(G)$ and $\mathscr{E}^*(G)$ as constructed in Exercises 3.7 and 3.8 are dual structures on E.

*3.10 Let $G = (V, E)$ be a graph, and let $\mathscr{E}$ be the collection of subsets A of E such that each component of (V, A) contains at most one cycle. Show that $\mathscr{E}$ is an independence structure on E. [Further properties of $\mathscr{E}$ will emerge later; see Exercise 4.16.]

3.11 Let $G = (V, E)$ be a graph, and let v, w be the endpoints of an edge e of G. *The contraction of G away from e* is the graph obtained from G by deleting e and 'identifying' v and w; i.e. replacing v and w by a single vertex which is an endpoint of precisely those edges (other than e) which originally had v or w as an endpoint. For example, the contraction of the left-hand graph below away from e is shown on the right.

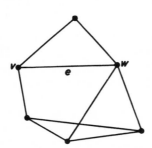
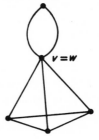

Let G' be a graph obtained from G by a sequence of contractions away from the distinct edges $e_1, \ldots, e_r$ of G and let $E' = \{e_1, \ldots, e_r\}$. Prove that

   (i) $\mathscr{E}(G') = \mathscr{E}(G)_{\otimes E \backslash E'}$
   (ii) $\mathscr{E}^*(G') = \mathscr{E}^*(G) | E \backslash E'$.

[Recall Exercise 2.18.]

3.12 Let $G = (V, E)$ be a connected graph. Show that $A \subseteq E$ is a disjoint union of cycles if and only if $|A \cap C|$ is even for every cutset C of G.

3.13 Let $G = (V, E)$ be a connected graph with $|E| = n$. Find the dimensions of the subspaces X and Y of $V_n = (GF(2))^n$ (defined in Section 3.2) and compare them with the ranks of $\mathscr{E}(G)$ and $\mathscr{E}^*(G)$. Show also that $X + Y = V_n$ if and only if there is no non-empty subset of E which is both a disjoint union of cycles and a disjoint union of cutsets.

3.14 Let G be the graph illustrated. Write down an incidence matrix of G. Confirm that each row corresponds to a disjoint union of cutsets of G, and express the vector corresponding to the cutset $\{e_2, e_3\}$ as a sum (over $GF(2)$) of rows of the incidence matrix in two different ways.

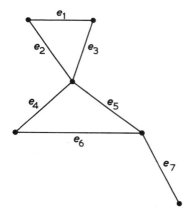

3.15 Let $G = (V, E)$ be a connected graph with $E = \{e_1, \ldots, e_n\}_{\neq}$, and let T be a spanning tree of G. Show that the vectors in $V_n = (GF(2))^n$ corresponding to the cycles defined in Exercise 3.2(i) span the subspace X, and that those corresponding to the cutsets defined in 3.2(ii) span Y. Give an example of a graph G for which these two collections of vectors together do not span the whole of V_n.

*3.16 Let $(E, \mathscr{E})$ be an independence space with $E = \{e_1, \ldots, e_n\}_{\neq}$, and let V_n be the vector space $(GF(2))^n$. Suppose that each subset of E is associated with a unique vector in V_n, in the way described in Section 3.2 (although here E is not necessarily interpreted as the edge-set of a graph). Prove that the following two properties are equivalent:

(i) for some positive integer s there exists an $s \times n$ matrix M with its rows in V_n and its columns labelled $e_1, \ldots, e_n$ such that

$A(\subseteq E)$ belongs to $\mathscr{E}$ if and only if the corresponding columns of M are distinct and form a linearly independent set over $GF(2)$;

(ii) every non-empty symmetric difference of circuits of $\mathscr{E}$ itself contains a circuit of $\mathscr{E}$.

3.17 Let $G = (V, E)$ be a connected graph with $|V| \geqslant 3$, and let $G^{\neq}$ be the graph obtained from G by deleting its loops and identifying repeated edges. Show that $(E, \mathscr{E}(G))$ is a circuit space (see Exercise 2.22) if and only if $G^{\neq}$ consists of a single cycle.

3.18 Let $G = (V, E)$ be a graph and ρ the rank function of its cycle structure. Prove that E can be partitioned into k forests if and only if

$$k\rho(A) \geqslant |A|$$

for every 'connected' set $A \subseteq E$; i.e. one which, together with its endpoints, forms a connected graph.

3.19 Let $G = (V, E)$ be the 'wheel' shown in the figure, consisting of $n + 1$ vertices and $2n$ edges ($n \geqslant 2$), and let ρ, be the rank function of its cycle structure. Show that

$$2\rho(A) \geqslant |A|$$

for each connected $A \subseteq E$. Demonstrate an actual partitioning of the wheel into two forests.

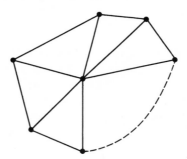

3.20 Let $G = (V, E)$ be the complete graph K_n, and let ρ be the rank function of its cycle structure. Show that

$$\frac{n}{2}\rho(A) \geqslant |A|$$

for each connected $A \subseteq E$. Deduce that the edge-set E of K_n can be partitioned into $\{n/2\}$ forests, and show further that this number cannot be reduced. Demonstrate a partitioning of K_5 into three forests. (Here, and in Exercise 3.21, $\{x\}$ denotes the smallest integer which is not less than x.)

3.21 Let $G = (V, E)$ be the complete bipartite graph $K_{r,s}$, and let ρ be the rank function of its cycle structure. Show that

$$\frac{rs}{r+s-1}\rho(A) \geqslant |A|$$

for each connected $A \subseteq E$. Deduce that the edge-set E of $K_{r,s}$ can be partitioned into $\{rs/(r+s-1)\}$ forests, and show further that this number cannot be reduced. Demonstrate a partitioning of $K_{3,5}$ into three forests.

3.22 Find necessary and sufficient conditions for a connected graph to be such that its edge-set can be partitioned into k sets, none of which disconnects the graph. Show that, if $G = (V, E)$ is the complete graph K_5, then

$$2c(A) + |A| \leqslant 12$$

for each $A \subseteq E$. Demonstrate a partitioning of this E into k sets, none of which disconnects the graph, where k is as small as possible.

3.23 Show that a connected graph $G = (V, E)$ can have its edge-set partitioned into k sets $(k > 1)$ such that the union of any $k - 1$ of them is a spanning tree of G if and only if

$$k(|V| - 1) = (k - 1)|E|$$

and, for each $A \subset E$,

$$k(|V| - c(A)) \geqslant (k - 1)|A|.$$

CHAPTER FOUR

Transversal spaces

4.1 Hall's theorem and its generalization

Let $\mathfrak{A} = (A_1, \ldots, A_n)$ be a family of subsets of a given set E. A subset $\{x_1, \ldots, x_n\}_{\neq}$ of E such that $x_i \in A_i$ for each i ($1 \leqslant i \leqslant n$) is called a *transversal* of $\mathfrak{A}$. Of course, not every family possesses a transversal. For example, the family

$$(\{a, b, c\}, \{a, b\}, \{d, e\}, \{a, c\})$$

(where a, b, c, d, e are assumed distinct) has several transversals, one being $\{a, b, c, d\}$, with (for instance) $a \in \{a, b, c\}$, $b \in \{a, b\}$, $d \in \{d, e\}$ and $c \in \{a, c\}$; whereas the family

$$(\{a, b\}, \{a, b\}, \{a\})$$

has none. A *partial transversal* of $\mathfrak{A}$ of length l is a transversal of a subfamily of l sets of $\mathfrak{A}$. So, for example, the latter family above has a partial transversal $\{a, b\}$ of length 2, this being a transversal of $(\{a, b\}, \{a\})$.

We may express the above ideas rather more formally: thus, a set $X (\subseteq E)$ is a transversal of $\mathfrak{A} = (A_1, \ldots, A_n)$ if there exists a bijective mapping $\sigma : X \to \{1, \ldots, n\}$ such that $x \in A_{\sigma(x)}$ for each $x \in X$, and X is a partial transversal of $\mathfrak{A}$ if there exists an injective mapping $\sigma : X \to \{1, \ldots, n\}$ such that $x \in A_{\sigma(x)}$ for each $x \in X$. For convenience, we shall admit partial transversals of zero length.

We note in passing that a transversal of a family of sets provides this family with a 'system of distinct representatives,' usually in more than one way. Thus, as above, $\{a, b, c, d\}$ is a transversal of the family

$$(\{a, b, c\}, \{a, b\}, \{d, e\}, \{a, c\})$$

and, since $a \in \{a, b, c\}$, $b \in \{a, b\}$, $d \in \{d, e\}$, $c \in \{a, c\}$, we may, for instance, take a, b, d, c as respective 'representatives' of the sets $\{a, b, c\}$, $\{a, b\}$, $\{d, e\}$ and $\{a, c\}$ in this order; equally well, we may take c, b, d, a as their respective representatives.

Clearly, if $\mathfrak{A} = (A_1, \ldots, A_n)$ possesses a transversal $\{x_1, \ldots, x_n\}_{\neq}$ with $x_i \in A_i$ for each i, then

$$\left| \bigcup_{i \in I'} A_i \right| \geq |\{x_i : i \in I'\}| = |I'|$$

for each $I' \subseteq \{1, \ldots, n\}$. In his now classic theorem (Theorem 4.1 below), Philip Hall proved that these necessary conditions

$$\left| \bigcup_{i \in I'} A_i \right| \geq |I'| \quad \forall I' \subseteq \{1, \ldots, n\}$$

are also sufficient for the existence of a transversal of $\mathfrak{A}$.

Let us write $\{1, \ldots, n\} = I$. We may evidently assume, with no essential loss of generality and whenever convenient, that I and E are disjoint, and then we may associate with the family $\mathfrak{A}$ a bipartite graph $G = (I, \Delta, E)$, where Δ is defined by the equation:

$$\Delta = \{ie : e \in A_i\}.$$

By way of example, the graphs associated with the two particular families just described are illustrated below:

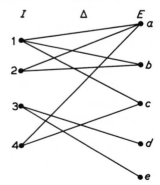

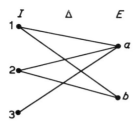

The conditions noted above for $\mathfrak{A}$ to possess a transversal may now be stated in the form

$$|\Delta(I')| \geqslant |I'| \quad \forall I' \subseteq I,$$

since, in agreement with Chapter 2,

$$\Delta(I') = \{e : ie \in \Delta \text{ for some } i \in I'\}$$

$$= \{e : e \in A_i \text{ for some } i \in I'\} = \bigcup_{i \in I'} A_i.$$

(On occasions, we shall feel free to adopt the convenient notation $\Delta(I')$ for $\bigcup_{i \in I'} A_i$ without making explicit reference to the underlying graph $G = (I, \Delta, E)$.)

Now let us suppose that an independence structure $\mathscr{E}$, with rank function ρ, is imposed on E. A transversal of $\mathfrak{A}$ which is a member of $\mathscr{E}$ will be called an *independent transversal* of $\mathfrak{A}$. Again, if $\mathfrak{A} = (A_1, \ldots, A_n)$ possesses an independent transversal $\{x_1, \ldots, x_n\}_{\neq}$ with $x_i \in A_i$ for each i, then

$$\rho\left(\bigcup_{i \in I'} A_i\right) \geqslant \rho(\{x_i : i \in I'\}) = |I'|$$

for each $I' \subseteq \{1, \ldots, n\}$; and once more these obvious necessary conditions for the existence of an independent transversal of $\mathfrak{A}$ turn out to be also sufficient. Evidently Hall's theorem is the special case of this more general result when $\mathscr{E}$ is the universal structure on E (and $\rho(A) = |A|$ for each $A \subseteq E$). The observant reader will have recognized that, in fact, we have already established this general result in Theorem 2.15, since $\mathfrak{A}$ has an independent transversal if and only if there exists a matching in $G = (I, \Delta, E)$ between I and an independent subset of E; and such a matching exists if and only if

$$\rho(\Delta(I')) \geqslant |I'| \quad \forall I' \subseteq I.$$

We shall not, however, dismiss Hall's theorem so lightly. Its place in the history of transversal theory justifies our spending some time on it. Hall's original proof was not particularly easy and was replaced by simpler arguments by later writers. One of the simplest of these is essentially similar to our proof of Theorem 2.15, and another is algorithmic and provides an actual construction of a transversal. Although this latter proof is far from being the shortest one, we describe it here because it is constructive and is of interest in its own right.

Theorem 4.1. (Hall's theorem.) *The family* $\mathfrak{A} = (A_1, \ldots, A_n)$ *of subsets of* E *possesses a transversal if and only if*

$$\left| \bigcup_{i \in I'} A_i \right| \geqslant |I'| \quad \forall I' \subseteq \{1, \ldots, n\}.$$

Proof (and algorithm). Since the necessity of the conditions has been established above, we now confirm that they are also sufficient. Suppose, then, that $|\Delta(I')| \geqslant |I'|$ for every $I' \subseteq \{1, \ldots, n\}$. Then, in particular, $|A_1| = |\Delta(1)| \geqslant |\{1\}| = 1$ and so $A_1 \neq \phi$. Hence the subfamily (A_1) of $\mathfrak{A}$ has a transversal. We shall establish by induction on m that $(A_1, \ldots, A_m)$ has a transversal for $1 \leqslant m \leqslant n$. So suppose that $1 \leqslant m < n$, that the result is true for m and, without loss of generality, write $E = \{x_1, \ldots, x_N\}_{\neq}$, where $x_1 \in A_1, \ldots, x_m \in A_m$. Our aim is now to find a transversal of $(A_1, \ldots, A_{m+1})$. If $A_{m+1} \nsubseteq \{x_1, \ldots, x_m\}$, then of course there exists an element of A_{m+1}, distinct from $x_1, \ldots, x_m$, which will serve to complete a transversal of $(A_1, \ldots, A_{m+1})$ as required. On the other hand, if $A_{m+1} \subseteq \{x_1, \ldots, x_m\}$, then we carry out the following process:

I. Let $E_0 = A_{m+1}$.

II. Given $E_t \subseteq \{x_1, \ldots, x_m\}$, denote by $I_t (\subseteq I)$ the set of suffixes of the xs in E_t, and form the set $E_{t+1} = \Delta(I_t)$.

III. Either $E_{t+1} \subseteq \{x_1, \ldots, x_m\}$, in which case we re-apply II with $t+1$ in place of t, or $E_{t+1} \nsubseteq \{x_1, \ldots, x_m\}$, in which case we stop.

Observe that, since $x_i \in A_i$ for each $i \in \{1, \ldots, m\}$, it follows that $E_t \subseteq \bigcup_{i \in I_t} A_i = \Delta(I_t) = E_{t+1}$ for each t until the process is stopped. In addition, since $A_{m+1} = E_0 \subseteq E_1 \subseteq \ldots \subseteq E_{t+1}$, it follows that

$$|E_{t+1}| = |\Delta(I_t) \cup A_{m+1}| = |\Delta(I_t \cup \{m+1\})|$$
$$\geqslant |I_t \cup \{m+1\}| > |I_t| = |E_t|,$$

and so $E_t \subseteq E_{t+1}$. Hence $E_0 \subseteq E_1 \subseteq E_2 \subseteq \ldots \subseteq \{x_1, \ldots, x_m\}$, and it follows that the above process must terminate, with $E_0, \ldots, E_{T+1}$ say, where $E_T = \Delta(I_{T-1}) \subseteq \{x_1, \ldots, x_m\}$ and $E_{T+1} = \Delta(I_T) \nsubseteq \{x_1, \ldots, x_m\}$.

Now there exists $b \in \Delta(I_T) \backslash \{x_1, \ldots, x_m\}$, and so $b \in A_{i_T}$ for some $i_T \in I_T \backslash I_{T-1}$; we shall use b to 'represent' A_{i_T} (in the transversal of $(A_1, \ldots, A_{m+1})$ to be constructed).

Next, $x_{i_T} \in E_T \backslash E_{T-1} = \Delta(I_{T-1}) \backslash \Delta(I_{T-2})$, and so $x_{i_T} \in A_{i_{T-1}}$ for some $i_{T-1} \in I_{T-1} \backslash I_{T-2}$; we shall use x_{i_T} to represent $A_{i_{T-1}}$.

. . . . .

Next, $x_{i_2} \in E_2 \backslash E_1 = \Delta(I_1) \backslash \Delta(I_0)$, and so $x_{i_2} \in A_{i_1}$ for some $i_1 \in I_1 \backslash I_0$; we shall use x_{i_2} to represent A_{i_1}.

At the last stage, $x_{i_1} \in E_1 \backslash E_0) = \Delta(I_0) \backslash E_0$, and so $x_{i_1} \in A_{i_0}$ for some $i_0 \in I_0$; we shall use x_{i_1} to represent A_{i_0} and x_{i_0} to represent $A_{m+1} (= E_0)$.

We finally observe that (since $x_i \in A_i$ for $1 \leqslant i \leqslant m$) the set $\{x_i : i \in \{1, \ldots, m\} \backslash \{i_0, \ldots, i_T\}\}$, together with $x_{i_0} (\in A_{m+1})$, $x_{i_1} (\in A_{i_0})$, $\ldots, x_{i_T} (\in A_{i_{T-1}})$, $b (\in A_{i_T})$, clearly provides a transversal $\{x_1, \ldots, x_m, b\}_{\neq}$ of $(A_1, \ldots, A_{m+1})$. Hence we have exhibited such a transversal; and it follows by induction that $(A_1, \ldots, A_n)$ itself has a transversal. $\square$

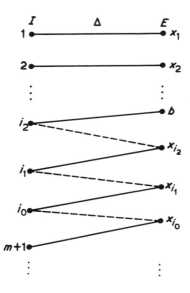

For comparison with Section 4.2 we represent the inductive step in the above proof (in the case $T = 2$) by the preceding diagram in the graph $G = (I, \Delta, E)$ associated with $\mathfrak{A}$. (In this diagram we show an 'alternating path' starting at b and ending at $m + 1$, in which the heavy lines join elements of E to the suffixes of the sets which they finally represent.) The essential feature of the construction is that we have introduced an element $b \notin \{x_1, \ldots, x_m\}$, and have thus 'freed' x_{i_0} (which originally represented A_{i_0}) so that it may represent A_{m+1} instead.

We illustrate the algorithm by an example. Of course, in situations with a small number of sets it is usually very easy to pick out a transversal (when one exists); our example is included merely to show the above construction at work.

Example

Let $\mathfrak{A} = (A_1, A_2, A_3, A_4, A_5)$ be the family of subsets of $\{x_1, x_2, x_3, x_4, x_5, x_6\}_{\neq}$, where $A_1 = \{x_1, x_2, x_3\}$, $A_2 = \{x_2, x_3, x_5, x_6\}$, $A_3 = \{x_1, x_3\}$, $A_4 = \{x_3, x_4\} = A_5$. We seek a transversal of $\mathfrak{A}$. To start with, we may choose

$$x_1 \text{ to represent } A_1,$$
$$x_2 \text{ to represent } A_2,$$
$$x_3 \text{ to represent } A_3,$$
$$x_4 \text{ to represent } A_4;$$

but then $A_5 \subseteq \{x_1, x_2, x_3, x_4\}$, and so we apply the algorithmic process of the above proof (with $m = 4$) to give

$$E_0 = A_5 = \{x_3, x_4\},$$
$$E_1 = \Delta(\{3, 4\}) = A_3 \cup A_4 = \{x_1, x_3, x_4\},$$
$$E_2 = \Delta(\{1, 3, 4\}) = A_1 \cup A_3 \cup A_4 = \{x_1, x_2, x_3, x_4\},$$
$$E_3 = \Delta(\{1, 2, 3, 4\}) = A_1 \cup A_2 \cup A_3 \cup A_4$$
$$= \{x_1, x_2, x_3, x_4, x_5, x_6\}$$
$$\nsubseteq \{x_1, x_2, x_3, x_4\}.$$

Again, in the notation above, $T = 2$ and

$$(b =) \; x_5 \in E_3 \backslash \{x_1, x_2, x_3, x_4\}, \; x_5 \in A_2,$$

$$x_2 \in E_2 \backslash E_1, \; x_2 \in A_1,$$

$$x_1 \in E_1 \backslash E_0, \; x_1 \in A_3,$$

$$x_3 \in E_0 = A_5;$$

which produces the elements $x_1, x_2, x_3, (x_4,) \, x_5$ as representatives of the sets $A_3, A_1, A_5, (A_4,) \, A_2$, respectively. (The fact that a transversal of $\mathfrak{A}$ has been found incidentally confirms that the conditions of Theorem 4.1 are satisfied for $\mathfrak{A}$.)

We insert here, for future use, a few very simple corollaries to Hall's theorem. Further extensions of a much deeper nature appear in Section 4.4.

Corollary 4.2. *Let $A_1, \ldots, A_n$ be subsets of a set E, and let $r_1, \ldots, r_n$ be positive integers. Then there exists a set X which can be partitioned into $X_1 \cup \ldots \cup X_n$, with $|X_i| = r_i$ and $X_i \subseteq A_i$ for each i, if and only if*

$$\left| \bigcup_{i \in I'} A_i \right| \geqslant \sum_{i \in I'} r_i \quad \forall \; I' \subseteq \{1, \ldots, n\}$$

Proof. Let

$$\mathfrak{A} = (\underbrace{A_1, \ldots, A_1}_{r_1}, \underbrace{A_2, \ldots, A_2}_{r_2}, \ldots, \underbrace{A_n, \ldots, A_n}_{r_n})$$

be a family consisting of r_i copies of A_i for $1 \leqslant i \leqslant n$. Now it is readily seen that a set X exists with the required properties if and only if this new family $\mathfrak{A}$ has a transversal. By Hall's theorem, this happens if and only if, for each $s(\leqslant \sum_{i=1}^{n} r_i)$, the union of the sets in any subfamily of s members of $\mathfrak{A}$ has cardinality at least s. Let us, then, look at a typical subfamily of s members of $\mathfrak{A}$; it will consist of (say) s_i copies of A_i, where $0 \leqslant s_i \leqslant r_i$ for $1 \leqslant i \leqslant n$ and $\sum_{i=1}^{n} s_i = s$. If we further write $I' = \{i : s_i > 0\}$, then $\sum_{i \in I'} s_i = s$; and a set X exists as required if and only if

$$\left| \bigcup_{i \in I'} A_i \right| \geqslant \sum_{i \in I'} s_i \qquad \begin{array}{l} \forall \ I' \subseteq \{1, \ldots, n\} \text{ and} \\ \forall \ \text{integers } s_i \text{ with } 0 < s_i \leqslant r_i. \end{array}$$

Evidently it is only necessary to check these conditions in the extreme case when each $s_i = r_i$; and so finally the required X exists if and only if

$$\left| \bigcup_{i \in I'} A_i \right| \geqslant \sum_{i \in I'} r_i \qquad \forall \ I' \subseteq \{1, \ldots, n\}. \qquad \square$$

Corollary 4.3. *The family* $\mathfrak{A} = (A_1, \ldots, A_n)$ *of subsets of* E *possesses a partial transversal of specified length* l (> 0) *if and only if*

$$\left| \bigcup_{i \in I'} A_i \right| \geqslant |I'| - n + l \qquad \forall \ I' \subseteq \{1, \ldots, n\}.$$

Proof. If $\mathfrak{A}$ possesses such a partial transversal, or if the given conditions hold, certainly $l \leqslant n$. Let, then, D be a set of $n - l$ elements disjoint from E and let $A_i' = A_i \cup D$ for each $i \in \{1, \ldots, n\}$. We verify at once that $\mathfrak{A}$ has a partial transversal of length l if and only if the family $(A_1', \ldots, A_n')$ has a transversal. By Theorem 4.1, this is so if and only if

$$\left| \bigcup_{i \in I'} A_i' \right| \geqslant |I'| \qquad \forall \ I' \subseteq \{1, \ldots, n\};$$

or, since these conditions are automatically satisfied if $I' = \phi$, if and only if

$$\left| \bigcup_{i \in I'} A_i' \right| \geqslant |I'| \qquad \forall \ I' \neq \phi, \ I' \subseteq \{1, \ldots, n\}.$$

But, for $I' \neq \phi$,

$$\left| \bigcup_{i \in I'} A_i' \right| = \left| \bigcup_{i \in I'} A_i \right| + |D|,$$

and so $(A_1', \ldots, A_n')$ has a transversal if and only if

$$\left| \bigcup_{i \in I'} A_i \right| \geqslant |I'| - |D| \qquad \forall \ I' \subseteq \{1, \ldots, n\};$$

and the desired result follows since $|D| = n - l$. $\qquad \square$

Our next two corollaries specify under what conditions a given set is a partial transversal of $\mathfrak{A}$.

Corollary 4.4. *Let $M(\subseteq E)$ be given. Then M is a partial transversal of* $\mathfrak{A} = (A_1, \ldots, A_n)$ *if and only if*

$$\left| \left(\bigcup_{i \in I'} A_i \right) \cap M \right| \geq |I'| + |M| - n \quad \forall\, I' \subseteq \{1, \ldots, n\}.$$

Proof. Clearly M is a partial transversal of $\mathfrak{A} = (A_1, \ldots, A_n)$ if and only if $(A_1 \cap M, \ldots, A_n \cap M)$ has a partial transversal of length $|M|$. By the previous corollary, this happens if and only if

$$\left| \bigcup_{i \in I'} (A_i \cap M) \right| = \left| \left(\bigcup_{i \in I'} A_i \right) \cap M \right| \geq |I'| + |M| - n \quad \forall\, I' \subseteq \{1, \ldots, n\}. \qquad \square$$

In the definition of a bipartite graph there is a symmetry between the two disjoint sets of vertices, and we shall now exploit this symmetry to obtain from Theorem 4.1 a second set of necessary and sufficient conditions for a given set M to be a partial transversal of a given family. Before deriving this corollary, we illustrate this symmetry (and the way it is to be used) in an example.

Example

Let $E = \{x_1, x_2, x_3, x_4, x_5\}_{\neq}$. The family $\mathfrak{A} = (A_1, A_2, A_3, A_4)$ where $A_1 = \{x_1, x_2, x_3\}$, $A_2 = \{x_3, x_4\}$, $A_3 = \{x_2, x_3, x_4\}$, $A_4 = \{x_2, x_4, x_5\}$, can clearly be represented by a bipartite graph, as in (1) below.

(1)

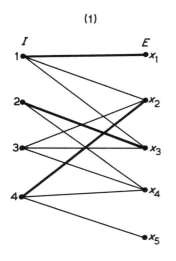

(2)

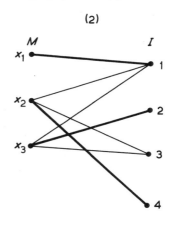

Restricting attention to $M = \{x_1, x_2, x_3\} \subseteq E$ and looking at this 'in reverse' we obtain the diagram (2). We observe that (2) is precisely the usual bipartite graph which is associated with the family $\mathfrak{F} = (F_{x_i} : x_i \in M)$, where $F_{x_1} = \{1\}$, $F_{x_2} = \{1, 3, 4\}$, $F_{x_3} = \{1, 2, 3\}$; and, in general $F_{x_j} = \{i \in I : x_j \in A_i\}$. Now the fact that M is a partial transversal of $\mathfrak{A}$ is illustrated in (1), and it is clear that this corresponds, in (2), to the fact that $\mathfrak{F}$ has a transversal.

In general, let $G = (I, \Delta, E)$, with $I = \{1, \ldots, n\}$, be the usual graph associated with the family $\mathfrak{A} = (A_1, \ldots, A_n)$ of subsets of E and let $M \subseteq E$. Define $\mathfrak{F} = (F_m : m \in M)$ by the equations $F_m = \{i \in I : m \in A_i\}$ for each $m \in M$. The example has illustrated that M is a partial transversal of $\mathfrak{A}$ if and only if there is a matching in G between M and a subset of I which, in turn, is the case if and only if $\mathfrak{F}$ has a transversal.

Corollary 4.5. *Let M ($\subseteq E$) be given. Then M is a partial transversal of $\mathfrak{A} = (A_1, \ldots, A_n)$ if and only if*

$$|\{i \in \{1, \ldots, n\} : M' \cap A_i \neq \phi\}| \geqslant |M'| \quad \forall \, M' \subseteq M.$$

Proof. For each $m \in M$, let $F_m = \{i \in \{1, \ldots, n\} : m \in A_i\}$ and write $\mathfrak{F} = (F_m : m \in M)$. Then it follows, as just observed, that M is a partial transversal of $\mathfrak{A}$ if and only if the family $\mathfrak{F}$ has a transversal. By Theorem 4.1, this is so if and only if

$$\left| \bigcup_{m \in M'} F_m \right| \geqslant |M'| \quad \forall \, M' \subseteq M.$$

But

$$\bigcup_{m \in M'} F_m = \{i \in \{1, \ldots, n\} : m \in A_i \text{ for some } m \in M'\}$$

$$= \{i \in \{1, \ldots, n\} : M' \cap A_i \neq \phi\};$$

and the result follows. $\qquad\square$

We have already remarked upon the generalization of Theorem 4.1 in the context of independence theory (namely Theorem 2.15). For convenience we formally restate it here in terms of sets and families.

Theorem 4.6. (Rado's theorem.) *Let $(E, \mathscr{E})$ be an independence space with rank function ρ. Then the family $\mathfrak{A} = (A_1, \ldots, A_n)$ of subsets of E*

possesses an independent transversal if and only if

$$\rho\left(\bigcup_{i \in I'} A_i\right) \geqslant |I'| \quad \forall\, I' \subseteq \{1, \ldots, n\}. \qquad \square$$

4.2 The partial transversals of a family of sets

We continue to use the notation of Section 4.1, and we write $\mathfrak{A} = (A_1, \ldots, A_n)$ (or equivalently, $(A_i : i \in I)$, where $I = \{1, \ldots, n\}$) for the family of sets under consideration. Further, we denote by $\mathscr{E}(\mathfrak{A})$ the collection of partial transversals of $\mathfrak{A}$, and speak of it as the *transversal structure* of $\mathfrak{A}$, and $(E, \mathscr{E}(\mathfrak{A}))$ as the associated *transversal space*. Most of this section is devoted to just one fundamental result, considered in some detail, namely that the transversal structure of $\mathfrak{A}$ is an independence structure. As in the case of Hall's theorem, we have already given a proof of this result, in a disguised and generalized form (in Theorem 2.16). However, in order to gain some further insight into the result, we shall describe several other arguments by which it can be established. The first proof we give is constructive, from basic principles, and it uses an algorithmic process which in some ways resembles our proof of Hall's theorem; the second proof is much shorter because it relies on the established theory of sums of independence spaces; and the third proof uses vector spaces and links up with the theory of 'linear representability' of independence spaces, which is the subject of Chapter 5.

Theorem 4.7. *The transversal structure $\mathscr{E}(\mathfrak{A})$ of a family $\mathfrak{A}$ of subsets of E is an independence structure on E.*

First proof. We observe that only the replacement property I(2) needs verification; and we recall that, as remarked earlier, the property that $X (\subseteq E)$ is a partial transversal of $\mathfrak{A}$ is equivalent to the existence of a matching between X and a subset of I in the usual bipartite graph $G = (I, \Delta, E)$ associated with $\mathfrak{A}$. Thus, to establish I(2), let X_1, $X_2 (\subseteq E)$ be partial transversals of $\mathfrak{A}$ with $|X_1| + 1 = |X_2|$, and let I_1, $I_2 (\subseteq I)$ be such that there exist matchings $\Delta_1, \Delta_2 (\subseteq \Delta)$ between I_1 and X_1, and I_2 and X_2, respectively. We shall find it convenient below to write $x_i = \Delta_1(i)$ for each $i \in I_1$. We note that this is, of course, equivalent to the statement $\Delta_1(x_i) = i$, and that it implies that $x_i \in A_i$ for $i \in I_1$. Now every member i of $I_2 \backslash I_1$ determines an 'alternating path' in G with consecutive vertices

$$i \quad, \quad \Delta_2(i), \quad \Delta_1\Delta_2(i), \quad \Delta_2\Delta_1\Delta_2(i), \ldots$$
$$(\in I) \quad (\in E) \quad (\in I) \quad (\in E)$$

(where $\Delta_1\Delta_2(i)$, for instance, is written instead of the more clumsy $\Delta_1(\Delta_2(i))$), continued for as long as possible, and eventually forced to end either after an even number of steps (in $I_1 \backslash I_2$) or after an odd number of steps (in $X_2 \backslash X_1$). Clearly, if one such $i \in I_2 \backslash I_1$ gives an alternating path leading to $i' \in I_1 \backslash I_2$, then the alternating path

$$i', \quad \Delta_1(i'), \quad \Delta_2\Delta_1(i'), \quad \Delta_1\Delta_2\Delta_1(i'), \ldots,$$

continued for as long as possible, leads back to i. Thus, if different members of $I_2 \backslash I_1$ lead to $I_1 \backslash I_2$, then they lead to different endpoints there. Since

$$|I_1 \backslash I_2| = |I_1| - |I_1 \cap I_2| = |X_1| - |I_1 \cap I_2| < |X_2| - |I_1 \cap I_2|$$
$$= |I_2| - |I_1 \cap I_2| = |I_2 \backslash I_1|,$$

it follows that not all the alternating paths starting in $I_2 \backslash I_1$ can end in $I_1 \backslash I_2$; so there exists at any rate one such path which ends in $X_2 \backslash X_1$. Let the consecutive vertices of such a path be

$$\bar{i} \quad, \quad x_{i_0} = \Delta_2(\bar{i}), \quad i_0 = \Delta_1(x_{i_0}), \quad x_{i_1} = \Delta_2(i_0),$$
$$(\in I_2 \backslash I_1) \quad (\in X_1 \cap X_2) \quad (\in I_1 \cap I_2) \quad (\in X_1 \cap X_2)$$

$$\cdots \quad x_{i_T} = \Delta_2(i_{T-1}), \quad i_T = \Delta_1(x_{i_T}), \quad b = \Delta_2(i_T).$$
$$(\in X_1 \cap X_2) \quad (\in I_1 \cap I_2) \quad (\in X_2 \backslash X_1)$$

In order to establish I(2), all that remains now is to note that $x_i \in A_i$ for $i \in I_1 \backslash \{i_0, \ldots, i_T\}$, that $x_{i_0} \in A_{\bar{i}}$, $x_{i_1} \in A_{i_0}, \ldots, x_{i_T} \in A_{i_{T-1}}$, $b \in A_{i_T}$; and hence that $X_1 \cup \{b\}$, where $b \in X_2 \backslash X_1$, is a partial transversal of $\mathfrak{A}$. $\qquad \square$

The process in this proof is illustrated opposite (for $T = 2$) and can be compared with the process in our proof of Hall's theorem; indeed, we have chosen the notation to highlight the similarities.

Second proof. For each i, $1 \leqslant i \leqslant n$, let $\mathscr{E}_i$ consist of ϕ and every singleton subset of A_i. Then it is evident that each $\mathscr{E}_i$ is an independence structure on E. But $X = \{x_1, \ldots, x_m\}_{\neq}$ is a partial transversal of $\mathfrak{A}$ if and only if there exist distinct elements $j_1, \ldots, j_m \in \{1, \ldots, n\}$ such that $x_i \in A_{j_i}$ (or $\{x_i\} \mathscr{E}_{j_i}$) for $1 \leqslant i \leqslant m$; i.e. if and only if $X = X_1 \cup \ldots \cup X_n$, where each X_i (either a singleton or ϕ) is in $\mathscr{E}_i$ for $1 \leqslant i \leqslant n$. Hence X is a partial transversal of $\mathfrak{A}$ if and only if

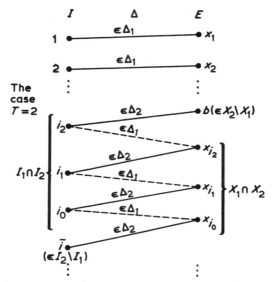

The case $T = 2$

it is an independent set in the sum $\mathscr{E}_1 + \ldots + \mathscr{E}_n$. Therefore $\mathscr{E}(\mathfrak{A}) = \mathscr{E}_1 + \ldots + \mathscr{E}_n$ and, by the extension of Theorem 2.17 to n summands, it follows that $\mathscr{E}(\mathfrak{A})$ is an independence structure on E. ☐

As we have already mentioned, our third proof will make use of vector spaces; and we now devote a few lines to some preliminary reminders in case the ideas we employ seem unfamiliar at first. We recall that 'rational functions' may be formed with (say) real coefficients in any number of independent variables $z_1, \ldots, z_p$. Thus, for example,

$$\frac{z_1 z_3}{z_1^2 + 2z_2^2}, \quad \frac{z_2^2 - 3z_1}{z_1 z_2 + z_1 z_3}, \quad \frac{1}{z_1 z_2 + z_2^3}, \quad z_1^2 - z_2^2 z_3 \left(\text{or } \frac{z_1^2 - z_2^2 z_3}{1} \right)$$

are rational functions in z_1, z_2, z_3. Those with denominator 1, such as the last one above, are called polynomials. In general, a rational function is just a quotient of polynomials. Two polynomials are regarded as equal when their corresponding coefficients are all equal; or, equivalently, the only polynomial in $z_1, \ldots, z_p$ which is zero is the one in which every coefficient is zero. With the usual rules for equating, adding and multiplying rational functions, it is easy to check that the set of all rational functions in $z_1, \ldots, z_p$ with real coefficients forms a field. In the proof below we shall work exclusively within this field and so, for brevity, we denote it throughout simply by F. We remark also before proceeding to the proof that, in an $m \times m$ square matrix, a product of m elements, no two of which lie in the same row or column, will be referred to as a 'permutation product' of the matrix.

Third proof. Let $E = \{x_1, \ldots, x_k\}_{\neq}$ and define an $n \times k$ matrix M associated with $\mathfrak{A}$ by the following procedure. Suppose there are exactly p ordered pairs (i, j) for which $x_j \in A_i$ and let us fill the corresponding positions in M with independent variables $z_1, \ldots, z_p$, and every other position in M with zero. Then M is a matrix with elements in the field F. Now for $X = \{x_{j_1}, \ldots, x_{j_m}\}_{\neq} \subseteq E$, let M_X be the submatrix of M which uses the columns $j_1, \ldots, j_m$ of M. It is easy to see that X is a partial transversal of $\mathfrak{A}$ if and only if M_X contains an $m \times m$ submatrix N with at least one non-zero permutation product. Since the determinant of N is an alternating sum of permutation products, it is clear that if $\det N \neq 0$ then certainly N has a non-zero permutation product. Conversely, if N has a non-zero permutation product then, since this particular product appears nowhere else in the expansion of $\det N$, therefore $\det N \neq 0$. We conclude that X is a partial transversal of $\mathfrak{A}$ if and only if M_X has an $m \times m$ submatrix with non-zero determinant; i.e. if and only if the columns of M_X are distinct and form a linearly independent set over the field F. Since, by Theorem 1.1, linearly independent sets of vectors satisfy I(2), it follows that the partial transversals of $\mathfrak{A}$ satisfy I(2); and the result follows. $\qquad\square$

Example

Let $E = \{x_1, x_2, x_3, x_4, x_5\}_{\neq}$ and $\mathfrak{A} = (A_1, A_2, A_3, A_4)$, where

$$A_1 = \{x_1, x_3, x_4, x_5\}, \quad A_2 = \{x_1, x_3\}, \quad A_3 = \{x_2, x_3\},$$
$$A_4 = \{x_3, x_4, x_5\}.$$

$$M = \begin{pmatrix} z_1 & 0 & z_2 & z_3 & z_4 \\ z_5 & 0 & z_6 & 0 & 0 \\ 0 & z_7 & z_8 & 0 & 0 \\ 0 & 0 & z_9 & z_{10} & z_{11} \end{pmatrix}$$

Observe, for example, that $\{x_1, x_2, x_3\}$ is a partial transversal of $\mathfrak{A}$.

(Indeed, the underlined entries in M show that it is a transversal of (A_1, A_2, A_3).) Also

$$\det \begin{pmatrix} z_1 & 0 & z_2 \\ z_5 & 0 & z_6 \\ 0 & z_7 & z_8 \end{pmatrix} = -z_1 z_6 z_7 + z_2 z_5 z_7 \neq 0,$$

so that the columns 1, 2, 3 of M form a linearly independent set (over F).

It may already have occurred to the reader that different families of sets can very well have the same transversal structure. To give a very simple instance, let $\mathfrak{A} = (\{1, 2\}, \{1, 2\})$ and $\mathfrak{B} = (\{1\}, \{2\}, \{1, 2\})$. Then $\mathscr{E}(\mathfrak{A}) = \mathscr{E}(\mathfrak{B})$, each being the universal structure on $\{1, 2\}$. We observe that one of these two families contains just two sets, and that this is equal to the rank of its transversal structure. This situation is quite general as we see from the theorem below.

Theorem 4.8. *If $\mathscr{E}$ is a transversal structure of rank n, then there exists a family $\mathfrak{A}$ of n sets such that $\mathscr{E} = \mathscr{E}(\mathfrak{A})$.*

Proof. Let $\mathscr{E}$, of rank n, be the collection of partial transversals of the family $(A_i : i \in J)$ where, of course, $|J| \geqslant n$. Let $\mathfrak{A} = (A_i : i \in I)$, with $I \subseteq J$ and $|I| = n$, be any subfamily of this which possesses a transversal; and write $\mathfrak{B} = (A_i : i \in J \backslash I)$. Now if X is in $\mathscr{E}$, then X is a partial transversal of $(A_i : i \in J)$ and so $X = X_1 \cup X_2$, where $X_1 \in \mathscr{E}(\mathfrak{A})$ and $X_2 \in \mathscr{E}(\mathfrak{B})$. Therefore $X_1 \subseteq Y$ for some basis Y of $\mathscr{E}(\mathfrak{A})$. Hence X_2 is a partial transversal of $\mathfrak{B}$ and Y is a transversal of $\mathfrak{A}$, and so $Y \cup X_2$ is a partial transversal of $(A_i : i \in J)$. Therefore $Y \cup X_2 \in \mathscr{E}$ and $|Y \cup X_2| \leqslant n = |Y|$. It follows that $X_2 \subseteq Y$, and thus that $X = X_1 \cup X_2 \subseteq Y$; and $X \in \mathscr{E}(\mathfrak{A})$. We conclude that $\mathscr{E} \subseteq \mathscr{E}(\mathfrak{A})$ and, since clearly $\mathscr{E}(\mathfrak{A}) \subseteq \mathscr{E}$, it follows that $\mathscr{E}$ is precisely the transversal structure of the family $\mathfrak{A}$ (of n sets). $\square$

4.3 Duals of transversal structures

We now set out to answer the natural question: what are the duals of transversal structures like? In order to do so, we shall need to make use of 'directed graphs'. Let, then, $G = (V, E)$ be a graph and associate a 'direction' in an arbitrary fashion with each of its edges: this

produces a *directed graph*; and any one of these we shall denote by
$G = (V, E)$. For example, from the graph

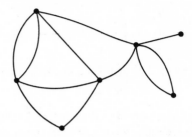

one of the many directed graphs which we may produce is

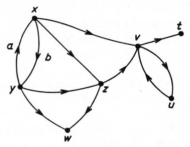

where, in a natural way, we indicate with an arrow the direction of
each edge. In what follows, loops will be irrelevant and therefore
excluded, as also will multiple edges with the *same* directions (but
not those such as a and b in the second figure). Clearly, then, in a
directed graph, the edges may be identified with *ordered* pairs of
vertices; for instance, the edge a with the pair (y, x), and we shall write
simply $a = (y, x)$. Our understanding of a path in a directed graph is
similar to that in a graph, as described in Section 1.2, with the natural
restriction that the edges proceed in the same direction (so that, for
example, there is a path from x to y to z to w above). So, in general, a
path in G will be a sequence

$$v_1, \ (v_1, v_2), \ v_2, \dots, v_n, \ (v_n, v_{n+1}), \ v_{n+1},$$

where $v_1, \dots, v_{n+1}$ are distinct vertices of G and $(v_1, v_2), \dots,$
(v_n, v_{n+1}) are edges of G. The *initial* and *terminal* vertices of this path
are v_1, v_{n+1}, respectively. We admit the case $n = 0$, when the path is
'degenerate'. A set of paths in G is *disjoint* if no two of them have a
vertex in common. Now let $G = (V, E)$ be a directed graph and let X,

$Y \subseteq V$. We say that X is *linked* *to* Y (in G) if there exists a set of disjoint paths in G whose set of initial vertices is X and whose set of terminal vertices is Y. So, in the above example, $\{x, z\}$ is linked to $\{u, w\}$. (We also allow an empty set of paths to link empty sets.) In theorem 4.9 below, we establish our first connection between linkages and transversals. For its statement we need one further item of notation. For each $v \in V$, we shall write

$$S_v = \{v\} \cup \{w : (v, w) \in E\}$$

and shall call S_v the *star* at v; further, for $X \subseteq V$, we shall write

$$\mathfrak{S}_X = (S_v : v \in X).$$

We observe that $\mathfrak{S}_X$ is a family of subsets of V (with a transversal X).

Theorem 4.9. *Let $G = (V, E)$ be a directed graph, and let $X, Y \subseteq V$. Then X can be linked to Y in G if and only if $V \backslash X$ is a transversal of the family $\mathfrak{S}_{V \backslash Y}$.*

Proof. The situation is only non-trivial if $\phi \neq X$, $Y \subset V$. First, then, suppose that X is linked to Y by the collection of disjoint paths $P_x(x \in X)$, and define a mapping $\alpha : V \backslash X \to V$ by the rule:

$$\alpha(u) = \begin{cases} v & \text{if the edge } (v, u) \text{ belongs to some } P_x \\ u & \text{otherwise.} \end{cases}$$

Since the paths P_x $(x \in X)$ are disjoint it is clear that α is well defined and is injective. It is also easy to check that no $\alpha(u)$ is a terminal vertex of any P_x, and hence that $\alpha(u) \in V \backslash Y$ for each $u \in V \backslash X$. Since $|X| = |Y|$ and $|V \backslash X| = |V \backslash Y|$, it follows that $\alpha : V \backslash X \to V \backslash Y$ is a bijection. Further, since $u \in S_{\alpha(u)}$ for each $u \in V \backslash X$, then $V \backslash X$ is a transversal of $\mathfrak{S}_{V \backslash Y}$.

Conversely, suppose that $V \backslash X$ is a transversal of $\mathfrak{S}_{V \backslash Y}$, so that there exist a bijection $\alpha : V \backslash X \to V \backslash Y$ with $u \in S_{\alpha(u)}$ for each $u \in V \backslash X$. Before proceeding to construct a collection of disjoint paths linking X to Y, we note that if $y \in Y \backslash X$ ($\subseteq V \backslash X$), then $\alpha(y)$ is defined and, since $y \in S_{\alpha(y)}$ and $y \neq \alpha(y)$ (since $\alpha(y) \notin Y$), it follows that $(\alpha(y), y)$ is an edge of G. If, then, $\alpha(y) \in V \backslash X$, then $\alpha(\alpha(y)) = \alpha^2(y)$ is defined and $(\alpha^2(y), \alpha(y))$ is an edge of G; and so on. Now, if $y, y' \in Y \backslash X$ and $\alpha^m(y) = \alpha^n(y')$ with $m \leq n$ say, then the injectivity of α ensures that $m = n$ and $y = y'$. (For if $m < n$, then $y = \alpha^{n-m}(y') \in V \backslash Y$; which is impossible.) In particular, given $y \in Y \backslash X$, the sequence

$$y, \quad \alpha(y), \quad \alpha^2(y), \ldots$$

(in which the successor of a term is defined so long as that term is in $V \backslash X$) consists of distinct vertices of G. Also, since $V \backslash X$ is finite, there exists a (first) positive integer M such that $\alpha^M(y) \notin V \backslash X$. So now we are in a position to construct a set of disjoint paths linking X to Y. For each $y \in Y \backslash X$, let P_y be the path with consecutive vertices

$$\alpha^M(y), \quad \alpha^{M-1}(y), \ldots, \alpha(y), \quad y,$$

where M is chosen, as above, so that $\alpha^M(y) \in X$. Further, for each $y \in Y \cap X$, let P_y be the degenerate path with the single vertex y. Then the collection of paths $P_y (y \in Y)$ certainly has Y for its set of terminal vertices. The above comments ensure also that, if $y \neq y'$, then no $\alpha^m(y)$ can coincide with any $\alpha^n(y')$; and so the paths P_y are disjoint. Since their initial vertices belong to X, and since $|V \backslash X| = |V \backslash Y|$ implies that $|X| = |Y|$, the set of initial vertices is precisely X. Hence X is linked to Y in G as required.   □

Given $A, B \subseteq V$, where $G = (V, E)$ is again a directed graph, the collection of all subsets of A which can be linked to subsets of B is called a *gammoid*. In the case when $A = V$, the gammoid is said to be *strict*. Thus, in our earlier illustrated example of a directed graph, the collection

$$\{\phi, \{x\}, \{y\}, \{z\}, \{x, y\}, \{x, z\}, \{y, z\}\}$$

is a gammoid (it consists of just those subsets of $\{x, y, z\}$ which can be linked to subsets of $\{v, x, z\}$). Further, the collection

$$\{\phi, \{x\}, \{y\}, \{z\}, \{v\}, \{u\}, \{x, y\}, \{x, z\}, \{x, v\},$$
$$\{x, u\}, \{y, z\}, \{y, v\}, \{y, u\}, \{z, v\}, \{z, u\}, \{x, y, v\},$$
$$\{x, y, u\}, \{x, z, v\}, \{x, z, u\}, \{y, z, v\}, \{y, z, u\}\}$$

is a strict gammoid (it consists of just those subsets of V which can be linked to subsets of $\{v, x, z\}$). The reader may already have suspected that gammoids are independence structures; and we shall provide an elementary and direct proof of this fact at the end of the section. However, the strict gammoids form a special class of independence structures, as we now demonstrate.

Theorem 4.10. *Strict gammoids are precisely the duals of transversal structures.*

Proof. The theorem asserts first that, if $\mathscr{E}$ is a strict gammoid in $G = (V, E)$, then there is a family of subsets of V whose transversal structure has dual $\mathscr{E}$, and second that, if $\mathscr{E}$ is the dual of a transversal structure of a family of subsets of some set V, then there is a directed graph $G = (V, E)$ which has $\mathscr{E}$ as one of its strict gammoids.

So assume first that $\mathscr{E}$ is a strict gammoid in the directed graph $G = (V, E)$ and suppose that it is precisely the collection of sets linked to subsets of $B(\subseteq V)$, say. Then, by Theorem 4.9,

$$X \in (\mathscr{E}(\mathfrak{S}_{V \setminus B}))^* \Leftrightarrow X \subseteq Y \text{ for some basis } Y \text{ of } (\mathscr{E}(\mathfrak{S}_{V \setminus B}))^*$$
$$\Leftrightarrow X \subseteq Y \text{ for some basis } V \setminus Y \text{ of } \mathscr{E}(\mathfrak{S}_{V \setminus B})$$
$$\Leftrightarrow X \subseteq Y \text{ for some transversal } V \setminus Y \text{ of } \mathfrak{S}_{V \setminus B}$$
$$\Leftrightarrow X \subseteq Y \text{ for some } Y \text{ linked to } B$$
$$\Leftrightarrow X \text{ is linked to some subset of } B$$
$$\Leftrightarrow X \in \mathscr{E}.$$

Hence $\mathscr{E}$ is precisely the dual of the transversal structure associated with the family $\mathfrak{S}_{V \setminus B}$.

Conversely, let $\mathscr{E}$ be the dual of a transversal structure on a set V. Then, by Theorem 4.8, $\mathscr{E}^*$ is the transversal structure of some family $\mathfrak{A} = (A_1, \ldots, A_n)$ of subsets of V which possesses a transversal $\{v_1, \ldots, v_n\}$, say, with $v_i \in A_i$ for each i. Next, define a directed graph $G = (V, E)$ by the condition

$$E = \bigcup_{i=1}^{n} \{(v_i, w) : w \in A_i, w \neq v_i\}.$$

Then, evidently, for each i, A_i is the star S_{v_i}, and so if $B = V \setminus \{v_1, \ldots, v_n\}$ we see that $\mathscr{E}^* = \mathscr{E}(\mathfrak{A}) = \mathscr{E}(\mathfrak{S}_{V \setminus B})$. Therefore, precisely as above, $X \in (\mathscr{E}(\mathfrak{S}_{V \setminus B}))^* (= \mathscr{E})$ if and only if X is linked to some subset of B. Hence $\mathscr{E}$ is a strict gammoid, as required. $\quad\square$

Corollary 4.11. *Gammoids are independence structures.*

Proof. Since the dual of an independence structure is an independence structure, it follows at once from Theorem 4.10 that strict gammoids are independence structures. But gammoids in general are just the restrictions of strict gammoids and so, too, are independence structures. $\quad\square$

Thus we have now completely characterized the duals of transversal structures in terms of gammoids. In fact, the result that gammoids are independence structures can be established without reference to transversals. We will give one such proof based on the following linkage theorem, of interest in its own right.

Theorem 4.12. (The linkage theorem.) *Let* $G = (V, E)$ *be a directed graph and let* $X_0 \subseteq X \subseteq V$, $Y_0 \subseteq Y \subseteq V$ *be such that* X_0 *is linked to some subset of* Y *(in* G*) and some subset of* X *is linked to* Y_0. *Then* X^* *is linked to* Y^* *for some* X^*, Y^* *with* $X_0 \subseteq X^* \subseteq X$ *and* $Y_0 \subseteq Y^* \subseteq Y$.

Proof. Let X_0 be linked to a subset of Y by the set $\mathscr{P}$ of disjoint paths in G: we regard $\mathscr{P}$ as fixed throughout the discussion. Given any $P \in \mathscr{P}$ and any path Q in G with at least one vertex v in common with P, we can follow the route of P as far as v and then follow the route of Q. If this overall route has no repeated vertex, then it is a path in G and we shall denote it by $P|v|Q$. Further, given any set $\mathscr{Q}$ of paths we shall write $\Pi_{\mathscr{Q}}$ for the set of paths in $\mathscr{P} \cup \mathscr{Q}$ together with those arising in the above way as a combination of a $P \in \mathscr{P}$ and a $Q \in \mathscr{Q}$.

Now let $\mathscr{Q}$ be a set of disjoint paths in G whose initial vertices are in X and whose set of terminal vertices is Y_0, so chosen that $|\Pi_{\mathscr{Q}}|$ is as small as possible. We shall show that $\Pi_{\mathscr{Q}}$ must contain a set of disjoint paths linking X^* to Y^* for some sets X^*, Y^* with $X_0 \subseteq X^* \subseteq X$, $Y_0 \subseteq Y^* \subseteq Y$. For assume that this is not so and let $x_1, \ldots, x_r$ be precisely the vertices in X_0 which are not initial vertices of any member of $\mathscr{Q}$, and let $P_1, \ldots, P_r \in \mathscr{P}$ have $x_1, \ldots, x_r$, respectively, as their initial vertices. Since $\{P_1, \ldots, P_r\} \cup \mathscr{Q} (\subseteq \Pi_{\mathscr{Q}})$ does not contain a set of paths of the required type, it is clear that at least one of $P_1, \ldots, P_r$, say P_1, has a vertex in common with some member of $\mathscr{Q}$. Let v be the first vertex of P_1 which is used in any member of $\mathscr{Q}$, say Q, and let $\mathscr{Q}' = (\mathscr{Q} \backslash \{Q\}) \cup \{P_1|v|Q\}$. It is straightforward to check that the members of $\mathscr{Q}'$ are disjoint, that their initial vertices lie in X, their set of terminal vertices is Y_0, that $\Pi_{\mathscr{Q}'} \subseteq \Pi_{\mathscr{Q}}$, and that $Q \in \Pi_{\mathscr{Q}} \backslash \Pi_{\mathscr{Q}'}$. Therefore $|\Pi_{\mathscr{Q}'}| < |\Pi_{\mathscr{Q}}|$; which contradicts the choice of $\mathscr{Q}$. Hence $\Pi_{\mathscr{Q}}$ *does* contain a set of disjoint paths linking an X^* to a Y^* in the required way. $\qquad \square$

Corollary 4.13. *Gammoids are independence structures.*

Proof. Let $\mathscr{E}$ be the gammoid arising from sets of vertices A, B in the directed graph $G = (V, E)$. It is clear that if a set X is linked to a subset of B, then so is any subset of X. Hence it only remains to establish that $\mathscr{E}$ satisfies I(2). So let X_0, $X_1 \in \mathscr{E}$ with $|X_1| = |X_0| + 1$. We shall establish I(2) by finding an $X^* \in \mathscr{E}$ with $X_0 \subset X^* \subseteq X_0 \cup X_1$. To this end, assume that $X_1 (\subseteq A)$ is linked to $Y_0 \subseteq B$. Then X_0 is linked to a subset of B, and a subset of $X_0 \cup X_1$ is linked to $Y_0 \subseteq B$. So it follows from the linkage theorem that X^* is linked to Y^* for some X^*, Y^* with $X_0 \subseteq X^* \subseteq X_0 \cup X_1$ and $Y_0 \subseteq Y^* \subseteq B$. Hence $|X^*| = |Y^*| \geqslant |Y_0| = |X_1| > |X_0|$ and so $X_0 \subset X^* \subseteq X_0 \cup X_1 (\subseteq A)$. Also, clearly, $X^* \in \mathscr{E}$ as it is linked to $Y^* \subseteq B$. We have thus found a set X^* as required and established I(2).

Since it is easy to verify that transversal structures are gammoids in a bipartite graph (which has been directed in an obvious way), we have yet another proof that transversal structures are independence structures.

4.4 Extensions of Hall's Theorem

The extensions of Hall's theorem are numerous and varied and have, in recent years, been systematically studied; as a result 'transversal theory' has emerged as a subject in its own right. Here we shall look briefly at just two natural extensions of Hall's theorem. Purely set-theoretic proofs of both are available, but it is only when independence theory is brought to bear upon them that their real content is probably most clearly understood. In their turn they provide examples of the power of independence theory and, in particular, of the notion of transversal independence discussed in Section 4.2.

Theorem 4.14. *Let* $\mathfrak{A} = (A_1, \ldots, A_n)$ *be a family of subsets of E and let M be a given subset of E. Then* $\mathfrak{A}$ *possesses a transversal which contains M as a subset if and only if*

    (1) $\mathfrak{A}$ *possesses a transversal,*
and   (2) *M is a partial transversal of* $\mathfrak{A}$.

Proof. The conditions (1) and (2) are evidently necessary. Suppose, therefore, that they are satisfied and, as in Section 4.2, let $\mathscr{E}(\mathfrak{A})$ be the independence structure of partial transversals of $\mathfrak{A}$. Then (1) implies that the maximal members of $\mathscr{E}(\mathfrak{A})$ have cardinality n, and (2) that

$M \in \mathscr{E}(\mathfrak{A})$. It therefore follows that M is contained in a maximal member of $\mathscr{E}(\mathfrak{A})$, namely a transversal of $\mathfrak{A}$.   □

Hall's theorem and Corollary 4.4 immediately yield the following quantified form of Theorem 4.14.

Corollary 4.15. *The family* $\mathfrak{A} = (A_1, \ldots, A_n)$ *has a transversal containing* M *if and only if*

$$\left| \bigcup_{i \in I'} A_i \right| \geq |I'| \quad and \quad \left| \left(\bigcup_{i \in I'} A_i \right) \cap M \right| \geq |I'| + |M| - n$$

$$\forall I' \subseteq \{1, \ldots, n\}. \qquad \square$$

For our second extension of Hall's theorem we turn from the consideration of a single family of sets to that of two families. Let, then, $\mathfrak{A} = (A_1, \ldots, A_m)$ and $\mathfrak{B} = (B_1, \ldots, B_n)$ be families of subsets of E with say $n \geq m$. We seek conditions for the existence of a transversal of $\mathfrak{A}$ which is also a partial transversal of $\mathfrak{B}$, i.e. a common transversal of $\mathfrak{A}$ and of a subfamily of $\mathfrak{B}$. In the special case when $m = n$, our result will, of course, give conditions for a common transversal of $\mathfrak{A}$ and $\mathfrak{B}$.

Theorem 4.16. *The families* $\mathfrak{A} = (A_1, \ldots, A_m)$ *and* $\mathfrak{B} = (B_1, \ldots, B_n)$ *with* $n \geq m$ *are such that* $\mathfrak{A}$ *and a subfamily of* $\mathfrak{B}$ *possess a common transversal if and only if*

$$\left| \left(\bigcup_{i \in I'} A_i \right) \cap \left(\bigcup_{j \in J'} B_j \right) \right| \geq |I'| + |J'| - n$$

$$\forall I' \subseteq \{1, \ldots, m\}, \; J' \subseteq \{1, \ldots, n\}.$$

Proof. By Theorem 4.7, the transversal structure $\mathscr{E}(\mathfrak{B})$ of $\mathfrak{B}$ is an independence structure on E. We shall denote its rank function by $\rho_{\mathfrak{B}}$. Now, for $X \subseteq E$ and k a positive integer, X contains a partial transversal of $\mathfrak{B}$ of length k if and only if the family $(B_1 \cap X, \ldots, B_n \cap X)$ possesses a partial transversal of length k. By Corollary 4.3, this happens if and only if

$$\left| \bigcup_{j \in J'} (B_j \cap X) \right| - |J'| + n \geq k \quad \forall J' \subseteq \{1, \ldots, n\}.$$

Hence X contains a partial transversal of $\mathfrak{B}$ of length

$$\min_{J' \subseteq \{1, \ldots, n\}} \left\{ \left| \bigcup_{j \in J'} (B_j \cap X) \right| - |J'| + n \right\}$$

but none of any greater length. Therefore

$$\rho_{\mathfrak{B}}(X) = \min_{J' \subseteq \{1, \ldots, n\}} \left\{ \left| \bigcup_{j \in J'} (B_j \cap X) \right| - |J'| + n \right\}.$$

Now $\mathfrak{A}$ has a common transversal with a subfamily of $\mathfrak{B}$ if and only if $\mathfrak{A}$ has a transversal which belongs to $\mathscr{E}(\mathfrak{B})$. By Theorem 4.6, this is so if and only if

$$\rho_{\mathfrak{B}}\left(\bigcup_{i \in I'} A_i \right) \geq |I'| \quad \forall I' \subseteq \{1, \ldots, m\},$$

i.e. if and only if

$$\min_{J' \subseteq \{1, \ldots, n\}} \left\{ \left| \bigcup_{j \in J'} \left(B_j \cap \left(\bigcup_{i \in I'} A_i \right) \right) \right| - |J'| + n \right\} \geq |I'|$$

$$\forall I' \subseteq \{1, \ldots, m\},$$

i.e. if and only if

$$\left| \left(\bigcup_{i \in I'} A_i \right) \cap \left(\bigcup_{j \in J'} B_j \right) \right| \geq |I'| + |J'| - n$$

$$\forall I' \subseteq \{1, \ldots, m\}, J' \subseteq \{1, \ldots, n\}. \qquad \square$$

Corollary 4.17. *The families* $\mathfrak{A} = (A_1, \ldots, A_n)$ *and* $\mathfrak{B} = (B_1, \ldots, B_n)$ *possess a common transversal if and only if*

$$\left| \left(\bigcup_{i \in I'} A_i \right) \cap \left(\bigcup_{j \in J'} B_j \right) \right| \geq |I'| + |J'| - n \quad \forall I', J' \subseteq \{1, \ldots, n\}. \qquad \square$$

Our final theorem in this section uses the linkage result of Theorem 4.12. For brevity, we find it convenient to write $\mathfrak{A}' \subseteq \mathfrak{A}$ if $\mathfrak{A}'$ is a subfamily of $\mathfrak{A}$.

Theorem 4.18. *Let* $\mathfrak{A}$, $\mathfrak{B}$ *be families of subsets of E and let* $\mathfrak{A}_0 \subseteq \mathfrak{A}$, $\mathfrak{B}_0 \subseteq \mathfrak{B}$. *Then there exist families* $\mathfrak{A}^*$, $\mathfrak{B}^*$ *with* $\mathfrak{A}_0 \subseteq \mathfrak{A}^* \subseteq \mathfrak{A}$ *and* $\mathfrak{B}_0 \subseteq \mathfrak{B}^* \subseteq \mathfrak{B}$ *and such that* $\mathfrak{A}^*$, $\mathfrak{B}^*$ *have a common transversal if and only if* $\mathfrak{A}_0$ *has a common transversal with a subfamily of* $\mathfrak{B}$, *and* $\mathfrak{B}_0$ *has a common transversal with a subfamily of* $\mathfrak{A}$.

Proof. Only the sufficiency of the conditions requires proof; so let us assume that they are satisfied. Further, without loss of generality, we may suppose that $\mathfrak{A} = (A_i : i \in I)$ and $\mathfrak{B} = (B_j : j \in J)$ where I, J and E are pairwise disjoint. Let G be the directed graph whose vertex-set is $I \cup E \cup J$ and whose edge-set is

$$\{(i, e) : i \in I \text{ and } e \in A_i\} \cup \{(e, j) : j \in J \text{ and } e \in B_j\}.$$

Then it is easy to see that $I' \subseteq I$ is linked to $J' \subseteq J$ in G if and only if the subfamilies $(A_i : i \in I')$, and $(B_j : j \in J')$ of $\mathfrak{A}$ and $\mathfrak{B}$ have a common transversal; indeed, in that event, a common transversal is provided by the set of members of E used by the paths linking I' to J'. So an application of Theorem 4.12 to this graph at once yields the desired result.   □

Corollary 4.19. *Let the families $\mathfrak{A}_0 \subseteq \mathfrak{A}$, $\mathfrak{B}_0 \subseteq \mathfrak{B}$ be as in the theorem. Then there exist families $\mathfrak{A}^*$, $\mathfrak{B}^*$ with $\mathfrak{A}_0 \subseteq \mathfrak{A}^* \subseteq \mathfrak{A}$, $\mathfrak{B}_0 \subseteq \mathfrak{B}^* \subseteq \mathfrak{B}$ and possessing a common transversal if and only if*

$$\left| \left(\bigcup_{i \in I'} A_i \right) \cap \left(\bigcup_{j \in J'} B_j \right) \right| \geqslant |I'| + |J'| - |J|$$

$$\forall I' \subseteq I_0,\ J' \subseteq J$$

and

$$\left| \left(\bigcup_{i \in I'} A_i \right) \cap \left(\bigcup_{j \in J'} B_j \right) \right| \geqslant |I'| + |J'| - |I|$$

$$\forall I' \subseteq I,\ J' \subseteq J_0,$$

where I_0, I, J_0, J are the respective index sets of the families $\mathfrak{A}_0, \mathfrak{A}, \mathfrak{B}_0, \mathfrak{B}$.

Proof. This is an immediate consequence of Theorems 4.16 and 4.18.

□

4.5 Applications

For our first application we turn back to the theory of linkages discussed in Section 4.3 and deduce a celebrated graph-theoretic result of Menger. Let then $G = (V, E)$ be a directed graph and let $X, Y \subseteq V$. A set $S \subseteq V$ is said to *separate* Y from X in G if every path from X to Y contains a member of S. Now suppose further that $|X| = |Y| = m$. It is easy to see that if X is linked to Y then every such separating set S (which must contain at least one vertex from each path linking X to Y) must satisfy the condition $|S| \geqslant m$. Menger's theorem asserts that the converse of this statement, which is much less obvious, is nevertheless also true.

Theorem 4.20. (Menger's theorem.) *The sets of vertices X and Y with $|X| = |Y| = m$ are linked in G if every set S which separates Y from X is such that $|S| \geqslant m$.*

Proof. Let us assume that X is not linked to Y in G; we shall exhibit a set S which separates Y from X, with $|S| < m$. By Theorem 4.9, $V \backslash X$ is not a transversal of the family $\mathfrak{S}_{V \backslash Y}$. Therefore, with the notation as before, the family $(S_v \cap (V \backslash X) : v \in V \backslash Y)$ does not possess a transversal and so, by Hall's theorem, there exists $W \subseteq V \backslash Y$ such that

$$\left| \bigcup_{v \in W} (S_v \cap (V \backslash X)) \right| \left(= \left| \left(\bigcup_{v \in W} S_v \right) \backslash X \right| \right) < |W|.$$

Now let $S = (X \cup \bigcup_{v \in W} S_v) \backslash W$. We shall show that $|S| < m$ and that S separates Y from X. Note first that, as $W \subseteq \bigcup_{v \in W} S_v$,

$$|S| = |X| + \left| \left(\bigcup_{v \in W} S_v \right) \backslash X \right| - |W| < |X| = m.$$

Note also that any path from X to Y is either a path from $X \backslash W$ to Y or from W to Y. All paths of the former type certainly contain a member of $X \backslash W$, and hence of S. Further, since W is disjoint from Y, in any path from W to Y there is a first vertex not in W. By the definition of the star sets S_v, that vertex must be in $\bigcup_{v \in W} S_v$, and hence in S. Thus any path from X to Y contains a member of S; and S separates Y from X as required. □

Our version of Menger's theorem is slightly specialized: the general version is an easy deduction and appears as an exercise.

Hall's theorem and its extensions have many applications outside transversal theory, and we close this chapter by examining just a few of these. The next two theorems are concerned with Latin rectangles. Let r, s, n be positive integers with r, $s \leqslant n$. An $r \times s$ *Latin rectangle*, based on $1, \ldots, n$, is an $r \times s$ matrix whose entries are drawn from the set $\{1, \ldots, n\}$ and with the property that no two numbers in any row or in any column are equal. When $r = s = n$, it is called a *Latin square* of order n.

For $n = 4$, an example of a 2×4 Latin rectangle is

$$\begin{pmatrix} 1 & 2 & 3 & 4 \\ 3 & 4 & 2 & 1 \end{pmatrix}.$$

This can be built up to a Latin square of order 4, for example

$$\begin{pmatrix} 1 & 2 & 3 & 4 \\ 3 & 4 & 2 & 1 \\ 4 & 3 & 1 & 2 \\ 2 & 1 & 4 & 3 \end{pmatrix}$$

and the theorem below shows that this is no accidental situation.

Theorem 4.21. *Let r, n be positive integers with $r < n$. Every $r \times n$ Latin rectangle based on $1, \ldots, n$ can be built up to a Latin square of order n.*

Proof. Let R be an $r \times n$ Latin rectangle and, for $1 \leqslant i \leqslant n$, denote by A_i the set of numbers from $1, \ldots, n$ which do *not* occur in the ith column of R. Then $|A_1| = \ldots = |A_n| = n - r$ and each $x \in \{1, \ldots, n\}$ occurs in precisely $n - r$ A_is. Suppose that $I' \subseteq \{1, \ldots, n\}$, put $\bigcup_{i \in I'} A_i = \{x_1, \ldots, x_p\}_{\neq}$, and assume that each x_j occurs in just s_j of the sets $(A_i : i \in I')$. Then, by the above remark, each $s_j \leqslant n - r$ and so

$$(n - r)p \geqslant s_1 + \ldots + s_p = \sum_{i \in I'} |A_i| = (n - r)|I'|.$$

Hence

$$p = \left| \bigcup_{i \in I'} A_i \right| \geqslant |I'|$$

for each $I' \subseteq \{1, \ldots, n\}$, and it follows from Hall's theorem that $\mathfrak{A} = (A_1, \ldots, A_n)$ possesses a transversal. A moment's thought will show that this implies that a further row can be adjoined to R to form an $(r + 1) \times n$ Latin rectangle. This process may be repeated for as long as the number of rows is less than the number of columns, and eventually gives a Latin square of order n. $\qquad \square$

Not all Latin rectangles based on $1, \ldots, n$ can be extended to

Latin squares of order n. For example, the Latin rectangle

$$\begin{pmatrix} 1 & 3 & 4 & 5 \\ 3 & 5 & 1 & 2 \\ 5 & 1 & 3 & 4 \end{pmatrix}$$

based on 1, 2, 3, 4, 5 cannot be extended to a Latin square of order 5, for it is easy to see that in any 5×5 extension of this rectangle it would be impossible to place a 2 in each row without two of them occurring in the same column. So it seems that the reason why this Latin rectangle cannot be extended to a Latin square of order 5 is that the number 2 occurs too few times in the rectangle. This comment is made precise by the following result, which is an application of Theorem 4.14.

Theorem 4.22. *Let r, s, n be positive integers with $r, s < n$, and let R be an $r \times s$ Latin rectangle based on $1, \ldots, n$. For each k with $1 \leqslant k \leqslant n$, let $N(k)$ be the number of times which the integer k occurs in R. Then R can be extended to a Latin square of order n if and only if, for each k,*

$$N(k) \geqslant r + s - n.$$

Proof. First, suppose that R can be extended to a Latin square, partitioned in the following fashion:

$$\begin{array}{c c c} & {\scriptstyle s} & {\scriptstyle n-s} \\ {\scriptstyle r} & R & S \\ {\scriptstyle n-r} & T & U. \end{array}$$

Then, for $1 \leqslant k \leqslant n$, the integer k occurs $r - N(k)$ times in S, and clearly this cannot exceed the number of times k occurs in S and U together, namely $n - s$. Hence $r - N(k) \leqslant n - s$; and thus the necessity of the conditions is easily established.

Now suppose that these conditions are satisfied for all k. For $1 \leqslant i \leqslant r$, let us denote by A_i the set of numbers from $1, \ldots, n$ which do *not* occur in the ith row of R, and let $\mathfrak{A} = (A_1, \ldots, A_r)$. Clearly $|A_1| = \ldots = |A_r| = n - s$ and each integer k is in precisely $r - N(k)$ $(\leqslant n - s)$ of the A_is. Given $I' \subseteq \{1, \ldots, r\}$, let $\bigcup_{i \in I'} A_i = \{x_1, \ldots, x_p\}_{\neq}$. Then, rather as in the previous proof,

$$(n - s)p \geqslant (r - N(x_1)) + \ldots + (r - N(x_p))$$
$$\geqslant \sum_{i \in I'} |A_i| = (n - s)|I'|$$

and so

$$p = \left| \bigcup_{i \in I'} A_i \right| \geq |I'|$$

for each $I' \subseteq \{1, \ldots, r\}$. Hence it follows that $\mathfrak{A}$ *has a transversal.*

Next, let $M = \{k : N(k) - r + s - n\}$. (We shall show that M is a partial transversal of $\mathfrak{A}$.) Let $M' = \{m_1, \ldots, m_t\}_{\neq} \subseteq M$, and suppose that $M' \cap A_i \neq \phi$ for exactly h values of i. Then, in particular, $n - s = r - N(m_j)$ for each j with $1 \leq j \leq t$, and hence

$$(n-s)t = (r - N(m_1)) + \ldots + (r - N(m_t)) = \sum_{i=1}^{r} |M' \cap A_i| \leq (n-s)h$$

and, for each $M' \subseteq M$,

$$h = |\{i : 1 \leq i \leq r \quad \text{and} \quad M' \cap A_i \neq \phi\}| \geq |M'| = t.$$

It follows from Corollary 4.5 that M *is a partial transversal of* $\mathfrak{A}$.

The crucial step in the argument is now to invoke Theorem 4.14 in order to deduce from the italicized statements above that $\mathfrak{A}$ possesses a transversal which contains M as a subset. *Any* transversal of $\mathfrak{A}$ may clearly be used to form an $(s + 1)$th column to be adjoined to R so that the resulting $r \times (s + 1)$ matrix R' is a Latin rectangle. However, we must choose a transversal containing M in order to ensure that the conditions of the theorem are satisfied for R' and that the process of adjoining columns can continue. For, if $1 \leq k \leq n$, then either $N(k) > r + s - n$ (in which case k occurs at least $r + (s + 1) - n$ times in R') or $N(k) = r + s - n$ (in which case $k \in M$ and k is in the $(s + 1)$th column of R', so again k occurs at least $r + (s + 1) - n$ times in R'). So the above procedure may be repeated for as long as the number of columns is less than n; eventually to lead to an $r \times n$ Latin rectangle. By Theorem 4.21, this in turn may be extended to a Latin square of order n. ☐

Matrices with elements equal to 0 or 1 arise in many situations; we have already encountered them as incidence matrices associated with graphs. As our final application now, we use our results on common transversals to deduce an existence theorem for matrices with elements equal to 0 and 1 whose row- and column-sums lie within prescribed bounds.

Theorem 4.23. *For $1 \leqslant i \leqslant m$ and $1 \leqslant j \leqslant n$, let r_i, r_i^0, s_j, s_j^0 be integers with $0 \leqslant r_i^0 \leqslant r_i$ and $0 \leqslant s_j^0 \leqslant s_j$. Then there exists an $m \times n$ matrix $M = (m_{ij})$ of 0s and 1s such that, for all i, j,*

$$r_i^0 \leqslant i\text{th row-sum of } M \leqslant r_i$$

and

$$s_j^0 \leqslant j\text{th column-sum of } M \leqslant s_j$$

if and only if

$$|I'||J'| \geqslant \max \left\{ \sum_{i \in I'} r_i^0 - \sum_{j \notin J'} s_j, \sum_{j \in J'} s_j^0 - \sum_{i \notin I'} r_i \right\}$$

$$\forall I' \subseteq \{1, \ldots, m\}, J' \subseteq \{1, \ldots, n\}.$$

Proof. Define the sets E, A_i, B_j by the rules

$$E = \{(i, j) : 1 \leqslant i \leqslant m, \ 1 \leqslant j \leqslant n\},$$
$$A_i = \{(i, j) : 1 \leqslant j \leqslant n\} \quad (1 \leqslant i \leqslant m),$$
$$B_j = \{(i, j) : 1 \leqslant i \leqslant m\} \quad (1 \leqslant j \leqslant n).$$

Evidently, $A_1, \ldots, A_m$ and $B_1, \ldots, B_n$ are partitions of E and each $|A_i \cap B_j| = 1$. Let us now denote by $\mathfrak{A}_0$ and $\mathfrak{A}$ the families consisting respectively of r_i^0 and r_i copies of A_i for $1 \leqslant i \leqslant m$, and by $\mathfrak{B}_0$ and $\mathfrak{B}$ the families consisting respectively of s_j^0 and s_j copies of B_j for $1 \leqslant j \leqslant n$.

It is soon clear that a matrix M with the required properties exists if and only if there are families $\mathfrak{A}^*$, $\mathfrak{B}^*$ with $\mathfrak{A}_0 \subseteq \mathfrak{A}^* \subseteq \mathfrak{A}$, $\mathfrak{B}_0 \subseteq \mathfrak{B}^* \subseteq \mathfrak{B}$ which possess a common transversal. For, if such a matrix M exists, then a common transversal of the desired families $\mathfrak{A}^*$, $\mathfrak{B}^*$ is provided by the set $\{(i, j) : m_{ij} = 1\}$. Conversely, if $\mathfrak{A}^*$, $\mathfrak{B}^*$ exist as stated with a common transversal X for which each $A_i \cap X$ intersects precisely the sets B_j with $j \in J_i$, and each $B_j \cap X$ intersects precisely the sets A_i with $i \in I_j$, then $r_i^0 \leqslant |J_i| \leqslant r_i$ and $s_j^0 \leqslant |I_j| \leqslant s_j$ for all i, j. It follows that the matrix $M = (m_{ij})$ which we seek is defined by the condition that $m_{ij} = 1$ if and only if $A_i \cap B_j \cap X \neq \phi$.

Now, by Theorem 4.18, the above situation for $\mathfrak{A}^*$ and $\mathfrak{B}^*$ arises if and only if

(a) $\mathfrak{A}_0$ and a subfamily of $\mathfrak{B}$ possess a common transversal, and

(b) $\mathfrak{B}_0$ and a subfamily of $\mathfrak{A}$ possess a common transversal.

We saw in the proof of Corollary 4.2 how to apply Hall's theorem to a family with repeated members. A similar process applied to the conditions of Theorem 4.16 shows that (a) holds if and only if

$$\left| \left(\bigcup_{i \in I'} A_i \right) \cap \left(\bigcup_{j \in J'} B_j \right) \right| \quad \left(= |I'||J'| \right) \geqslant \sum_{i \in I'} r_i^0 + \sum_{j \in J'} s_j - \sum_{j=1}^{n} s_j$$

$$\forall I' \subseteq \{1, \ldots, m\}, J' \subseteq \{1, \ldots, n\}.$$

The condition (b) is dealt with similarly; and the required result follows.   □

Exercises

4.1 Consider a set of n boys each of whom is acquainted with a number of girls. Show that it is possible for each of the boys to marry a girl of his acquaintance if and only if for each r, $1 \leqslant r \leqslant n$, each set of r boys knows between them at least r girls. [In the popular literature, Hall's theorem is often referred to as the 'marriage theorem'.]

4.2 Let n, k be given positive integers with $k \geqslant n$ and let $M = (m_{ij})$ be an $n \times k$ matrix of 0s and 1s. Prove that M has a set of n elements equal to 1 no two of which lie in the same row or column if and only if, for each r with $1 \leqslant r \leqslant n$, every r rows of M contains elements equal to 1 lying in at least r columns.

4.3 Let $\mathfrak{A} = (A_1, \ldots, A_n)$ be a family of subsets of a set E. A *system of representatives* of $\mathfrak{A}$ is a family $(a_1, \ldots, a_n)$ of elements of E such that, for some ordering $j_1, \ldots, j_n$ of $1, \ldots, n$, $a_{j_i} \in A_i$ for each i. (Clearly, any family of non-empty sets has a system of representatives, and a family possesses a transversal if and only if it has a system of *distinct* representatives.) Now let $\mathfrak{A} = (A_1, \ldots, A_n)$, $\mathfrak{B} = (B_1, \ldots, B_n)$ be two families of subsets of E. Apply Hall's theorem to an appropriate family of subsets of $I = \{1, \ldots, n\}$ to show that $\mathfrak{A}$ and $\mathfrak{B}$ have a common system of representatives if and only if

$$\left| \left\{ j \in I : \left(\bigcup_{i \in I'} A_i \right) \cap B_j \neq \phi \right\} \right| \geqslant |I'| \quad \forall I' \subseteq I.$$

4.4 Let $\mathfrak{A} = (A_1, \ldots, A_n)$ and $\mathfrak{B} = (B_1, \ldots, B_n)$ be two families of subsets of E. Show that $\mathfrak{A}$ and $\mathfrak{B}$ have a common system of representatives (as in Exercise 4.3) if and only if

$$\left(\bigcup_{i \in I'} A_i \right) \cap \left(\bigcup_{j \in J'} B_j \right) \neq \phi$$

whenever $I', J' \subseteq \{1, \ldots, n\}$ are such that $|I'| + |J'| > n$.

4.5 Let $\mathfrak{A} = (A_1, \ldots, A_m)$ be a family of subsets of E, and let $\mathscr{E}$ be an independence structure on E with rank function ρ. Prove that $\mathfrak{A}$ has an independent partial transversal of length p if and only if

$$\rho \left(\bigcup_{i \in I'} A_i \right) \geq |I'| - m + p \quad \forall I' \subseteq \{1, \ldots, m\}.$$

4.6 Let $\mathfrak{A} = (A_1, \ldots, A_m)$ and $\mathfrak{B} = (B_1, \ldots, B_n)$ be families of subsets of a set E, and let $p \leq m, n$ be given. Show that $\mathfrak{A}$ and $\mathfrak{B}$ have a common partial transversal of length p if and only if

$$\left| \left(\bigcup_{i \in I'} A_i \right) \cap \left(\bigcup_{j \in J'} B_j \right) \right| \geq |I'| + |J'| - m - n + p$$

$$\forall I' \subseteq \{1, \ldots, m\}, J' \subseteq \{1, \ldots, n\}.$$

[Hint: use Exercise 4.5.]

4.7 Show that the common partial transversals of two families of subsets of E do not in general form an independence structure on E.

*4.8 A *Hamiltonian path* in a directed graph $G = (V, E)$ is a path which uses every vertex of G (precisely once). Let $V = \{v_0, \ldots, v_n\}_{\neq}$. Let $\mathscr{E}$ be the cycle structure of G (the undirected graph associated with G), and let $\bar{\mathscr{E}}$ be a 'copy' of $\mathscr{E}$ on $\bar{E}$. Let

$$A_i = \{e \in E : e = (x, v_i) \text{ for some } x \in V\} \quad (1 \leq i \leq n)$$

and

$$B_i = \{e \in E : e = (v_i, x) \text{ for some } x \in V\} \quad (0 \leq i \leq n - 1).$$

Show that G has a Hamiltonian path from v_0 to v_n if and only if $\mathfrak{A} = (A_1, \ldots, A_n)$ and $\mathfrak{B} = (B_0, \ldots, B_{n-1})$ have a common transversal in $\mathscr{E}$. [Necessary and sufficient conditions for an independent common transversal of two families have not yet been found; and this is one of the outstanding unsolved problems of the subject.]

4.9 Show that the conditions

$$\left|\left(\bigcup_{i \in I'} A_i\right) \cap \left(\bigcup_{j \in J'} B_j\right) \cap \left(\bigcup_{k \in K'} C_k\right)\right| \geqslant |I'| + |J'| + |K'| - 2n$$

$$\forall I', J', K' \subseteq \{1, \ldots, n\}$$

are necessary but not sufficient for the families $\mathfrak{A} = (A_1, \ldots, A_n)$, $\mathfrak{B} = (B_1, \ldots, B_n)$, $\mathfrak{C} = (C_1, \ldots, C_n)$ to have a common transversal.

4.10 Let $\mathfrak{A} = (A_1, \ldots, A_n)$, $\mathfrak{B} = (B_1, \ldots, B_n)$ be families of subsets of a set E, let $\mathscr{E}$ be the transversal structure of $\mathfrak{B}$, and let $M \subseteq E$. Use the rank function of the independence structure $\mathscr{E}(M) = \mathscr{E}|M \oplus \mathscr{E}_{\otimes E \setminus M}$ to show that $\mathfrak{A}$ and $\mathfrak{B}$ have a common transversal containing M if and only if

$$\left|\left(\bigcup_{i \in I'} A_i\right) \cap \left(\bigcup_{j \in J'} B_j\right)\right| + \left|\left(\bigcup_{i \in I'} A_i \cup \bigcup_{j \in J'} B_j\right) \cap M\right|$$

$$\geqslant |I'| + |J'| + |M| - n \qquad \forall I', J' \subseteq \{1, \ldots, n\}.$$

4.11 Let $\mathscr{E}$ be the transversal structure of the family $\mathfrak{A} = (A_1, \ldots, A_n)$. Show that each $E \setminus A_i$ is a flat in $\mathscr{E}$.

4.12 If $\mathscr{E}$ is a transversal structure of rank n, then any family $\mathfrak{A} = (A_1, \ldots, A_n)$ with $\mathscr{E} = \mathscr{E}(\mathfrak{A})$ is called a *presentation* of $\mathscr{E}$. Further, if no family $(B_1, \ldots, B_n)$ with $B_1 \subseteq A_1, \ldots, B_n \subseteq A_n$ apart from $\mathfrak{A}$ itself is a presentation of $\mathscr{E}$, then $\mathfrak{A}$ is a *minimal presentation* of $\mathscr{E}$. Show that a minimal presentation of a transversal structure can contain no repeated set.

*4.13 Let $\mathscr{E}$ be a transversal structure of rank n with presentation $(A_1, \ldots, A_n)$. Let B be a transversal of $(A_2, \ldots, A_n)$ such that, for any other transversal B' of $(A_2, \ldots, A_n)$, $|B' \cap A_1| \geqslant |B \cap A_1|$. Show that $(A_1 \setminus B, A_2, \ldots, A_n)$ is also a presentation of $\mathscr{E}$. Deduce that, if $(A_1, \ldots, A_n)$ is a minimal presentation of $\mathscr{E}$ (as in Exercise 4.12), then for each i there exists a transversal of $(A_1, \ldots, A_{i-1}, A_{i+1}, \ldots, A_n)$ contained in $E \setminus A_i$.

4.14 Let $\mathscr{E}$ be a transversal structure of rank n. Prove that the presentation $\mathfrak{A} = (A_1, \ldots, A_n)$ of $\mathscr{E}$ is minimal if and only if each A_i is a circuit of $\mathscr{E}^*$.

4.15 Use Exercise 4.14 to verify that $\mathfrak{A} = (\{1, 2\}, \{2, 3\}, \{4, 5\})$ is a minimal presentation of its transversal structure.

4.16 Let $G = (V, E)$ be a graph and let $\mathscr{E}$ be the collection of subsets A of E defined by the condition that the components of (V, A)

each contain at most one cycle (as in Exercise 3.10). Show that $A \in \mathscr{E}$ if and only if there exist $|A|$ distinct vertices of G such that each of them is the endpoint of a different member of A. Deduce that $\mathscr{E}$ is a transversal structure.

4.17 Let $G = (V, E)$ be the graph illustrated and let $\mathscr{E}$ be the independence structure defined in Exercise 4.16. Verify directly that $\mathscr{E} = \mathscr{E}(\mathfrak{A})$, where $\mathfrak{A} = (\{1, 4, 5\}, \{1, 2, 6\}, \{2, 3, 7\}, \{3, 4, 8\}, \{5, 6, 7, 8\})$, and show that $\mathfrak{A}$ is a minimal presentation of $\mathscr{E}$.

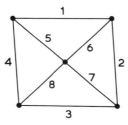

4.18 (General form of Menger's theorem.) Let $G = (V, E)$ be a directed graph, let $X, Y \subseteq V$ and let m be a positive integer. Prove that if every set separating Y from X in G contains at least m elements, then there exist $X^* \subseteq X$, $Y^* \subseteq Y$ with $|X^*| = |Y^*| = m$ and X^* linked to Y^* in G.

4.19 Let $\mathfrak{A} = (A_1, \ldots, A_n)$, $\mathfrak{B} = (B_{n+1}, \ldots, B_{2n})$ be families of subsets of $\{2n + 1, \ldots, m\}$ and let $G = (V, E)$ be the directed graph given by $V = \{1, \ldots, m\}$ and $E = \{(i, j) : j \in A_i \text{ or } i \in B_j\}$. Apply Menger's theorem to G to show that, if

$$\left| \left(\bigcup_{i \in I'} A_i \right) \cap \left(\bigcup_{j \in J'} B_j \right) \right| \geqslant |I'| + |J'| - n$$

$$\forall I' \subseteq \{1, \ldots, n\}, J' \subseteq \{n + 1, \ldots, 2n\},$$

then $\mathfrak{A}$ and $\mathfrak{B}$ have a common transversal.

*4.20 Given an undirected graph $G = (V, E)$ let $G = (V, E)$ be specified by $E = \{(u, v) : uv \in E\}$ (so $uv \in E$ implies that $(u, v), (v, u) \in E$). Any strict gammoid arising from a special directed graph of this kind is called an *undirected strict gammoid*. Prove that the strict gammoid consisting of the sets linked to subsets of $\{1, 3, 3''\}$ in the directed graph illustrated is not an undirected strict gammoid.

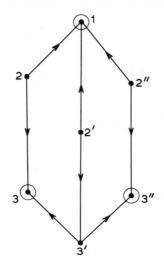

4.21 Extend the Latin rectangle

$$\begin{pmatrix} 1 & 2 & 3 & 4 & 5 \\ 3 & 4 & 2 & 5 & 1 \end{pmatrix}$$

to a 5 × 5 Latin square.

4.22 Find the possible values of i such that the Latin rectangle

$$\begin{pmatrix} 1 & 2 & 3 & 4 \\ 6 & 5 & 1 & 2 \\ 3 & 4 & 6 & 1 \\ 4 & 1 & 2 & i \end{pmatrix}$$

can be extended to a 6 × 6 Latin square. Carry through one such completion.

*4.23 Let $G = (V, E)$ be a graph and for each $M \subseteq E$ let $V(M)$ be the set of endpoints of members of M. We recall that a matching M in G is a set $M \subseteq E$ with $|V(M)| = 2|M|$; i.e. with no two of its endpoints in common. Define a collection $\mathscr{E}$ of subsets of V by the rule

$$\mathscr{E} = \{X \subseteq V : \text{there exists a matching } M \text{ with } X \subseteq V(M)\}.$$

Let $A \subseteq V$ be given. Use the following procedure to show that any two maximal subsets of A contained in $\mathscr{E}$ have the same cardinality and to deduce that $(V, \mathscr{E})$ is an independence space.

(i) Let X_1, X_2 be maximal subsets of A contained in $\mathscr{E}$ and let M_1, M_2 be matchings with $X_1 \subseteq V(M_1)$, $X_2 \subseteq V(M_2)$. Show that each component of the graph $G' = (V(M_1 \triangle M_2), (M_1 \triangle M_2))$ is a path or a cycle.

(ii) Show that $v \in X_1 \triangle X_2$ if and only if $v \in A$ and v is an endpoint of a path component of G'.

(iii) Prove that if $|X_2| > |X_1|$, then there exists a path component of G' with edge-set P and with endpoints v and w such that $v \in X_2 \backslash X_1$, $w \notin X_1 \backslash X_2$ and $\{v\} \cup X_1 \subseteq V(M_1 \triangle P)$.

(iv) Deduce that $|X_1| = |X_2|$, and that $(V, \mathscr{E})$ is an independence space.

4.24 A *matching structure* is a restriction of an independence structure arising from a graph in the way described in Exercise 4.23. Prove that a transversal structure is a matching structure. [In fact, the converse is also true, but is more difficult to prove.]

CHAPTER FIVE

Appendix on representability

The status of this chapter is a little different from that of the preceding ones and we draw attention to this by referring to it as an appendix. Whereas in Chapters 1 to 4 all results stated were also proved, here we find it desirable to include some results without proof. In spite of this, we still only touch upon the fringe of the representation problem.

5.1 Representability in general

We have met several special types of independence spaces in the earlier chapters: for instance subsets of a vector space, where independence was respectively linear and affine, the cycle and cutset spaces associated with a graph, and transversal spaces and their duals. If an independence space $(E, \mathscr{E})$ can be identified with a space of some special type, we shall regard that latter space as a 'model' for $(E, \mathscr{E})$ and say that it 'represents' $(E, \mathscr{E})$. It is interesting to question whether there actually exists one special type of space, arising naturally, which can provide a *universal* model for independence spaces.

Let us turn first to the cycle spaces. We readily see from either of the following examples that these spaces cannot provide us with such a universal model.

Examples

(1) Let $E = \{1, 2, 3, 4\}$ and let $\mathscr{E}$ be the truncation at 2 of the universal

structure on E. Then $(E, \mathscr{E})$ is not a cycle space. For, if G is a graph with four edges, then it is easy to see that it is impossible for every three edges of G to contain a cycle unless some pair of edges contains a cycle.

(2) (*The Fano geometry.*) Let $E = \{1, 2, 3, 4, 5, 6, 7\}$ and let $\mathscr{E}$ be the independence structure whose bases are all subsets of E of cardinality three with the exception of $\{1, 2, 6\}$, $\{2, 3, 4\}$, $\{1, 3, 5\}$, $\{1, 4, 7\}$, $\{2, 5, 7\}$, $\{3, 6, 7\}$ and $\{4, 5, 6\}$. Then $(E, \mathscr{E})$ is not a cycle space. For, as the reader can verify for himself, if each of the sets of edges $\{1, 2, 6\}$, $\{1, 4, 7\}$ and $\{2, 3, 4\}$ forms a cycle, then it is impossible for $\{1, 3, 5\}$ to be a cycle.

It is easily checked that all independence spaces of rank 1, and all independence spaces with most three elements are cycle spaces: so Example (1) above is a minimal 'non-cycle' space.

Since not every independence space is a cycle space, it follows at once by a consideration of duals that neither is every independence space a cutset space.

We have other simple grounds for ruling out the cycle and cutset spaces as universal models for independence spaces. For instance, these graphic spaces have the property that, if C, C' are distinct circuits, then $C \triangle C'$ always contains another circuit of the space, whereas this is not a general property of independence spaces (see, for example, Exercise 2.7).

In Chapter 1 we have referred to W. T. Tutte's now classic characterization theorems for graphic spaces; and here again we briefly draw the reader's attention to them. Let $(E, \mathscr{E})$ be an independence space. By a *minor* of $(E, \mathscr{E})$ we understand any independence space obtained from $(E, \mathscr{E})$ by a succession of restrictions and contractions. It is clear from Exercise 3.11 that any minor of a cycle space is a cycle space. Tutte has shown that an independence space is a cycle space if and only if it does not contain as a minor any one of five particular spaces: namely the truncation at two of the universal structure on a set of four members, the cutset spaces associated with $K_{3,3}$ and K_5, the Fano geometry and its dual. The interested reader must turn to a more advanced text for a proof of this result.

As is the case for the graphic spaces (i.e. the cycle and cutset spaces), so also the transversal spaces cannot provide a universal model, as the following example demonstrates.

Example

Let $E = \{1, 2, 3, 4, 5, 6\}$ and let $\mathscr{E}$ be the independence structure on E whose bases are all sets of cardinality 2 with the exception of $\{1, 2\}$, $\{3, 4\}$ and $\{5, 6\}$. Then $(E, \mathscr{E})$ is not a transversal space. For if it were transversal, then, by Theorem 4.8, it would have a presentation $\mathfrak{A} = (A_1, A_2)$ for some $A_1, A_2 \subseteq E$ with, say, $\{1, 2\} \subseteq A_1$ and $\{1, 2\} \cap A_2 = \phi$. Since $\{1, 3\} \in \mathscr{E}$ and $\{3, 4\} \notin \mathscr{E}$, it follows that $\{3, 4\} \subseteq A_2$ and $\{3, 4\} \cap A_1 = \phi$. Similarly, since $\{1, 5\} \in \mathscr{E}$ and $\{5, 6\} \notin \mathscr{E}$, it follows that $\{5, 6\} \subseteq A_2$ and $\{5, 6\} \cap A_1 = \phi$, and hence that $A_1 = \{1, 2\}$ and $A_2 = \{3, 4, 5, 6\}$. But then $\{4, 6\} \notin \mathscr{E}$; which is a contradiction. Hence $(E, \mathscr{E})$ is not a transversal space.

We shall also see in the exercises that the cycle space associated with the graph K_4 is not a transversal space. All other spaces with six or fewer elements, apart from these two, are known to be transversal. Of course, the duals of these two non-transversal spaces provide examples of independence spaces whose associated structures are not strict gammoids.

Again, as in the case of the graphic spaces, transversal spaces have special properties not shared by independence spaces in general. For example, if $A = \{a_1, \ldots, a_n\}_{\neq}$ and $B = \{b_1, \ldots, b_n\}_{\neq}$ are bases of a transversal space, then there exists a bijection $\theta : A \to B$ such that $(A \backslash \{a_i\}) \cup \{\theta(a_i)\}$ and $(B \backslash \{b_i\}) \cup \{\theta^{-1}(b_i)\}$ are also bases for each i, $1 \leqslant i \leqslant n$. This kind of simultaneous basis-exchange property, known as *base orderability*, is not held by independence spaces in general. The verification for the transversal spaces is left as an exercise; and also in the exercises there is an example of an independence space which does not have this property (and is hence not a transversal space).

Since abstract independence was motivated by linear independence, the most natural representations to look for are as subsets of a vector space. The rest of this chapter is devoted to a consideration of this particular representation problem.

5.2   Linear representability

The reader will recall that, in Theorem 3.13, we showed how to associate with each edge of a graph G a vertex belonging to the vector space $V = GF(2)^s$; for example by means of an incidence matrix.

Example

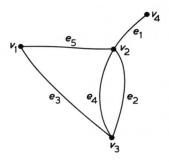

$$
\begin{array}{ccccc}
e_1 & e_2 & e_3 & e_4 & e_5 \\
\end{array}
$$

$$
\begin{array}{c}
v_1 \\ v_2 \\ v_3 \\ v_4
\end{array}
\begin{pmatrix} 0 \\ 1 \\ 0 \\ 1 \end{pmatrix},
\begin{pmatrix} 0 \\ 1 \\ 1 \\ 0 \end{pmatrix},
\begin{pmatrix} 1 \\ 0 \\ 1 \\ 0 \end{pmatrix},
\begin{pmatrix} 0 \\ 1 \\ 1 \\ 0 \end{pmatrix},
\begin{pmatrix} 1 \\ 1 \\ 0 \\ 0 \end{pmatrix}
\in (GF(2))^4.
$$

Let us now, in general, denote the vector associated with e_i by $\varphi(e_i)$. Then, as in Theorem 3.13, $\{e_{i_1}, \ldots, e_{i_r}\}_{\neq}$ is independent in the cycle space if and only if $\varphi(e_{i_1}), \ldots, \varphi(e_{i_r})$ are distinct members of V with $\{\varphi(e_{i_1}), \ldots, \varphi(e_{i_r})\}_{\neq}$ linearly independent. We have, in this way, 'represented' the cycle space of the graph as part of the vector space V, and this kind of representation motivates our definition of linear representability. An independence space $(E, \mathscr{E})$ with rank function ρ is called *linearly representable* if there exists a vector space V and a rank-preserving mapping $\varphi : E \to V$; i.e. a mapping $\varphi : E \to V$ for which the linear rank of $\varphi(A)$ is equal to $\rho(A)$ for every $A \subseteq E$. (We have not demanded that φ be injective and so, strictly, our definition is precisely in agreement with the notions of Section 5.1 only when the independence spaces considered have no dependent sets of cardinality 2.) We see from the next theorem that our definition describes exactly the sort of representation which we constructed formerly for the cycle spaces.

Theorem 5.1. *An independence space* $(E, \mathscr{E})$ *is linearly representable if and only if there exists a vector space V and a mapping $\varphi : E \to V$ such that, for $A \subseteq E$,*

$A \in \mathscr{E} \Leftrightarrow \varphi|A$ *(the restriction of φ to A) is injective and $\varphi(A)$ is linearly independent in V.*

Proof. Let $(E, \mathscr{E})$ have rank function ρ and, as in Chapter 1, for $U \subseteq V$ let dim U denote the linear rank of U. Then, for any mapping $\varphi : E \to V$ and for any $A \subseteq E$,

(i) dim $\varphi(A) \leqslant |\varphi(A)|$, with equality if and only if $\varphi(A)$ is linearly independent;

(ii) $\rho(A) \leqslant |A|$, with equality if and only if $A \in \mathscr{E}$;

(iii) $|\varphi(A)| \leqslant |A|$, with equality if and only if $\varphi|A$ is injective.

Therefore, if $\varphi : E \to V$ is rank preserving, then

$$A \in \mathscr{E} \Leftrightarrow |A| = \rho(A) \ (= \dim \varphi(A) \leqslant |\varphi(A)| \leqslant |A|)$$
$$\Leftrightarrow |A| = \dim \varphi(A) = |\varphi(A)|$$
$$\Leftrightarrow \varphi|A \text{ is injective and } \varphi(A) \text{ is linearly independent.}$$

Conversely, assume that φ satisfies the given conditions. Then, given $B \subseteq E$, let B have basis A and $\varphi(B)$ have basis $\varphi(A')$, where $A' \subseteq B$ and $\varphi|A'$ is injective. Then the given conditions apply to A and A' to give

$$\rho(B) = |A| = |\varphi(A)| = \dim \varphi(A) \leqslant \dim \varphi(B)$$

and

$$\rho(B) \geqslant \rho(A') = |A'| = |\varphi(A')| = \dim \varphi(B).$$

Thus $\rho(B) = \dim \varphi(B)$, and $\varphi : E \to V$ is rank-preserving. ☐

Corollary 5.2. *The cycle and cutset spaces of a graph are linearly representable in vector spaces over $GF(2)$.*

Proof. For connected graphs this is now immediate from Theorems 3.13, 3.15 and 5.1. The extension to general graphs is easy. ☐

Corollary 5.3. *Transversal spaces are linearly representable.*

Proof. Our third proof of Theorem 4.7 shows that any transversal space is linearly representable in a vector space over a function field. $\square$

Of course it is possible to represent a particular independence space in many different ways.

Example

In the graph illustrated in the previous example we represented the cycle space by means of a mapping $\varphi : E \to (GF(2))^4$ given by

$$
\begin{array}{ccccc}
\varphi(e_1) & \varphi(e_2) & \varphi(e_3) & \varphi(e_4) & \varphi(e_5) \\
\begin{pmatrix} 0 \\ 1 \\ 0 \\ 1 \end{pmatrix} &
\begin{pmatrix} 0 \\ 1 \\ 1 \\ 0 \end{pmatrix} &
\begin{pmatrix} 1 \\ 0 \\ 1 \\ 0 \end{pmatrix} &
\begin{pmatrix} 0 \\ 1 \\ 1 \\ 0 \end{pmatrix} &
\begin{pmatrix} 1 \\ 1 \\ 0 \\ 0 \end{pmatrix}
\end{array}
$$

Of the many other linear representations we now give one in $V = (GF(2))^3$, namely

$$
\begin{array}{ccccc}
\theta(e_1) & \theta(e_2) & \theta(e_3) & \theta(e_4) & \theta(e_5) \\
\begin{pmatrix} 1 \\ 0 \\ 0 \end{pmatrix} &
\begin{pmatrix} 0 \\ 1 \\ 0 \end{pmatrix} &
\begin{pmatrix} 0 \\ 0 \\ 1 \end{pmatrix} &
\begin{pmatrix} 0 \\ 1 \\ 0 \end{pmatrix} &
\begin{pmatrix} 0 \\ 1 \\ 1 \end{pmatrix}
\end{array}.
$$

This latter representation has two advantages: first, the dimension of the vector space coincides with the rank of the independence space (and hence it is the lowest dimension of a vector space possible) and, second, a chosen basis $\{e_1, e_2, e_3\}$ of $\mathscr{E}$ is mapped to the natural basis of V. Now we see that such a representation is always possible for linearly representable spaces.

Theorem 5.4. *Let* $(E, \mathscr{E})$ *be an independence space, with basis* $\{e_1, \ldots, e_r\}_{\neq}$, *which is linearly representable in a vector space over a field* F. *Then there is a rank-preserving mapping* $\theta : E \to F^r$ *of* $(E, \mathscr{E})$ *in which* $\theta(e_i) = (0, \ldots, \underset{(i)}{1}, \ldots, 0)$ *for* $1 \leqslant i \leqslant r$.

Proof. Let $\varphi : E \to V$ give rise to a linear representation of $\mathscr{E}$ in the vector space V over F. Then $\{\varphi(e_1), \ldots, \varphi(e_r)\}_{\neq}$ is linearly independent in V and, for any $e \in E$,

$$\varphi(e) = \lambda_1(e)\varphi(e_1) + \ldots + \lambda_r(e)\varphi(e_r)$$

for some uniquely-determined $\lambda_1(e), \ldots, \lambda_r(e) \in F$. Let $\theta : E \to F^r$ be defined by the rule

$$\theta(e) = (\lambda_1(e), \ldots, \lambda_r(e)) \qquad \forall \ e \in E.$$

Then $\theta(e_i) = (0, \ldots, \overset{(i)}{1}, \ldots, 0)$ for $1 \leqslant i \leqslant r$, as required; and it only remains to confirm that θ is rank-preserving. But, for $A \subseteq E$ and for any $\alpha_e \in F$ associated with $e \in A$,

$$\sum_{e \in A} \alpha_e \theta(e) = 0 \Leftrightarrow \sum_{e \in A} \alpha_e(\lambda_1(e), \ldots, \lambda_r(e)) = 0$$

$$\Leftrightarrow \sum_{e \in A} \alpha_e \lambda_1(e) = \ldots = \sum_{e \in A} \alpha_e \lambda_r(e) = 0$$

$$\Leftrightarrow \sum_{e \in A} \alpha_e(\lambda_1(e)\varphi(e_1) + \ldots + \lambda_r(e)\varphi(e_r)) = 0$$

$$\Leftrightarrow \sum_{e \in A} \alpha_e \varphi(e) = 0.$$

Hence the $\theta(e)$s satisfy precisely the same linear relations as the $\varphi(e)$s; and this is all that is required. Thus θ gives rise to the desired linear representation in F^r. ∎

Corollary 5.5. *Let $E = \{e_1, \ldots, e_n\}_{\neq}$ and let $(E, \mathscr{E})$ be an independence space, with basis $\{e_1, \ldots, e_r\}$, which is linearly representable in a vector space over a field F. Then there exists an $r \times n$ matrix M with entries in F whose first r columns form the identity matrix and such that the mapping which takes e_i to the ith column of M is rank preserving.*

Proof. If we take the special mapping θ constructed in Theorem 5.4 and form the matrix M with columns $\theta(e_1), \ldots, \theta(e_n)$, then M will be precisely as required. ∎

We conclude this section with a remark concerning affine representability. It is natural to call an independence space $(E, \mathscr{E})$ *affinely representable* if there exists a rank preserving mapping $\varphi : E \to V$, where V is a vector space and now the rank in V is affine rank. In fact, it can be shown that an independence space is affinely representable if and only if it is linearly representable, but the details need not concern us here. It is worth remarking that an affinely-representable independence space of rank $r (\geqslant 1)$ is so representable in a vector space of dimension $r - 1$.

5.3 Induced structures

We shall show now that duals, restrictions, contractions and truncations of linearly representable spaces are themselves linearly representable. The case of restrictions is trivial, for if $\varphi : E \to V$ gives rise to a representation of $(E, \mathscr{E})$, then the restriction of φ to A gives rise to a representation of the restriction $\mathscr{E} | A$.

Theorem 5.6. *Let F be a given field. If $(E, \mathscr{E})$ is linearly representable in a vector space over F, then so is its dual $(E, \mathscr{E}^*)$.*

Proof. The result is easy if the rank of $\mathscr{E}$ is zero. Let, then, $E = \{e_1, \ldots, e_n\}_{\neq}$, let $\mathscr{E}$ be of rank $r(> 0)$ and let M be any $r \times n$ matrix with entries in F such that the mapping which takes e_i to the ith column of M is rank-preserving. (The existence of such an M is ensured by Corollary 5.5) Then the equations

$$Mx = 0 \qquad (x \in F^n)$$

have just $n - r$ linearly independent solutions in F^n. Let us then denote by M^* an $n \times (n - r)$ matrix whose columns are just such a set of solutions. Now we make two claims about M^*.

(i) *A family of r columns of M is linearly related* (i.e. *contains a repeated column or forms a linearly dependent set of distinct columns*) *if and only if the complementary family of $n - r$ rows of M^* is linearly related.*

With no essential loss in generality we confirm this claim for the first r columns of M. We partition M and M^* as follows

$$
\begin{matrix} M & & M^* \\ r\begin{pmatrix} M_1 & M_2 \\ {}_r & {}_{n-r} \end{pmatrix} & \begin{pmatrix} M_3 \\ M_4 \end{pmatrix}\!\!\begin{matrix} {}^r \\ {}_{n-r} \end{matrix} & = 0. \\ & {}_{n-r} \end{matrix}
$$

Now observe that the family of the first r columns of M is linearly related

$\Leftrightarrow M_1$ is singular

$$\Leftrightarrow M\begin{pmatrix} \alpha_1 \\ \vdots \\ \alpha_r \\ 0 \\ \vdots \\ 0 \end{pmatrix} = 0 \text{ for some } \alpha_1, \ldots, \alpha_r \in F \text{ not all zero}$$

$$\Leftrightarrow \begin{pmatrix} \alpha_1 \\ \vdots \\ \alpha_r \\ 0 \\ \vdots \\ 0 \end{pmatrix} = \begin{pmatrix} M_3 \\ M_4 \end{pmatrix} \begin{pmatrix} \beta_1 \\ \vdots \\ \beta_{n-r} \end{pmatrix}$$ for some $\alpha_1, \ldots, \alpha_r \in F$, and some $\beta_1, \ldots, \beta_{n-r} \in F$ not all zero

(since the columns of M^* span the solution space of $M_x = 0$)

$$\Leftrightarrow M_4 \begin{pmatrix} \beta_1 \\ \vdots \\ \beta_{n-r} \end{pmatrix} = 0$$ for some $\beta_1, \ldots, \beta_{n-r} \in F$ not all zero

$\Leftrightarrow M_4$ is singular

$\Leftrightarrow$ The family of the last $n - r$ rows of M^* is linearly related.

(ii) *The mapping* $\varphi : E \to F^{n-r}$ *which takes* e_i *to the* ith *row of* M^* *is rank-preserving with respect to* $\mathscr{E}^*$.

Let us consider, without loss in generality, the set of elements $e_1, \ldots, e_s$ and write $\rho^*(\{e_1, \ldots, e_s\}) = k$, where ρ^* is the rank function of $\mathscr{E}^*$. Now the first s-rows of M^* have linear rank k. For

$\rho^*(\{e_1, \ldots, e_s\}) = k \Leftrightarrow$ there is a basis of $\mathscr{E}$ containing $s - k$ members of $\{e_1, \ldots, e_s\}$, but no basis containing fewer than this

$\Leftrightarrow$ there is a linearly independent set of r columns of M containing $s - k$ of the first s columns, but no such set containing fewer than this

$\Leftrightarrow$ there is a linearly independent set of $n - r$ rows of M^* containing k of the first s rows, but no such set containing more than this (Take complements and use (i))

$\Leftrightarrow$ the first s rows of M^*, namely $\varphi(e_1), \ldots, \varphi(e_s)$, have linear rank k.

Hence φ is rank-preserving and we have exhibited a linear representation of $(E \, \mathscr{E}^*)$. $\square$

Example. Let G be the graph shown overleaf. Then (as we have already seen) its cycle space can be represented by the columns (in $(GF(2))^3$) of the matrix

$$M = \begin{pmatrix} 1 & 0 & 0 & 0 & 0 \\ 0 & 1 & 0 & 1 & 1 \\ 0 & 0 & 1 & 0 & 1 \end{pmatrix}.$$

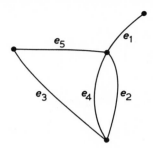

One possibility for the matrix M^* is then

$$M^* = \begin{pmatrix} 0 & 0 \\ 1 & 1 \\ 0 & 1 \\ 1 & 0 \\ 0 & 1 \end{pmatrix};$$

since its columns form a basis for the solution space of $Mx = 0$. It is, of course, simple to verify directly that the mapping $\varphi : \{e_1, e_2, e_3, e_4, e_5\} \rightarrow (GF(2))^2$, where $\varphi(e_i)$ is the ith row of M^*, provides a linear representation of the cutset space of G. (For example, $\{e_2, e_4\}$ is independent in $\mathscr{E}^*$ and $\{\varphi(e_2), \varphi(e_4)\}_{\neq}$ is linearly independent. On the other hand, $\{e_3, e_5\}$ is dependent in $\mathscr{E}^*$, and $\varphi(e_3) = \varphi(e_5)$.)

The reader will be able to show at once that, when M is taken in the 'standard form' of Corollary 5.5, namely

$$M = (I_r\ D),$$

then we may always choose

$$M^* = \begin{pmatrix} -D \\ I_{n-r} \end{pmatrix}.$$

This is illustrated in the above example (where, over $GF(2)$, $D = -D$).

Corollary 5.7. *Let F be a given field. If $(E, \mathscr{E})$ is linearly representable in a vector space over F, then so is any contraction of $(E, \mathscr{E})$.*

Proof. This is an immediate consequence of Theorem 5.6 since, as in Exercise 2.18, the contraction of $\mathscr{E}$ away from E' is given by $(\mathscr{E}^*|E\backslash E')^*$, and both duals and restrictions of linearly representable spaces are linearly representable (over the same field). $\square$

Corollary 5.8. *Gammoids are linearly representable.*

Proof. Gammoids are restrictions of duals of transversal spaces, and so the result is an immediate consequence of Theorem 5.6 and Corollary 5.3.   ☐

We have seen, therefore, that if an independence space is linearly representable in a vector space over a given field F, then so are its dual, contraction and restriction. We show next that the truncation of a linearly representable space is itself linearly representable, but in this case we may need to take a vector space over a larger field than F.

Theorem 5.9. *Let $(E, \mathscr{E})$ be a linearly representable independence space. Then any truncation of $(E, \mathscr{E})$ is also linearly representable.*

Proof. Let $(E, \mathscr{E})$ have rank r. The result is only non-trivial if $r \geqslant 2$. We consider in detail the case where $(E, \mathscr{E})$ is linearly representable in a vector space over a finite field F. Denote by $B_1, \ldots, B_p$ the independent sets of $(E, \mathscr{E})$ of cardinality $r - 1$. The remainder of our proof will require a representation of $(E, \mathscr{E})$ in a vector space over a field of cardinality exceeding p. Since (as in the exercises) F may be replaced by any of its extension fields, we may choose a vector space V of dimension r over such an extension F' for which $|F'| = q > p$ and for which there exists a rank-preserving mapping $\varphi : E \to V$. Then $\varphi(B_1), \ldots, \varphi(B_p)$ span hyperplanes, say $H_1, \ldots, H_p$ of V, and evidently

$$|H_1 \cup \ldots \cup H_p| \leqslant |H_1| + \ldots + |H_p| = p \cdot q^{r-1} < q^r = |V|.$$

So there exists a vector $v \in V$ with $v \notin H_1 \cup \ldots \cup H_p$ (and, in particular, $v \notin \varphi(E)$). Let $\bar{e} \notin E$ and consider the bipartite graph $(E \cup \{\bar{e}\}, \Delta, \varphi(E) \cup \{v\})$, where

$$\Delta = \{e\varphi(e) : e \in E\} \cup \{\bar{e}v\}.$$

By Theorem 2.16 the linearly independent subsets of $\varphi(E) \cup \{v\}$ induce an independence structure $\bar{\mathscr{E}}$ on $E \cup \{\bar{e}\}$, and we note that $(E \cup \{\bar{e}\}, \bar{\mathscr{E}})$ is linearly representable by means of the mapping $\Delta : E \cup \{\bar{e}\} \to V$. Hence, by Corollary 5.7, the contraction $(E, \bar{\mathscr{E}}_{\otimes E})$ (i.e. away from $\{\bar{e}\}$) is also linearly representable. We shall show that this contraction is precisely the truncation of $(E, \mathscr{E})$ at $r - 1$. First, given $B \in \mathscr{E}$ with $|B| \leqslant r - 1$, it follows that $B \cup \{\bar{e}\} \in \bar{\mathscr{E}}$ and hence that $B \in \bar{\mathscr{E}}_{\otimes E}$. Conversely, if $B \in \bar{\mathscr{E}}_{\otimes E}$, then $B \cup \{\bar{e}\} \in \bar{\mathscr{E}}$; thus $|B| \leqslant r - 1$ and $B \in \mathscr{E}$. The result follows for finite F.

The case of a vector space over an infinite field is not dissimilar, but relies on the non-trivial result that such a vector space is not equal to the union of any finite number of its hyperplanes.   □

5.4   Linear representability over specified fields

Having considered whether an independence space is linearly representable in some vector space, we wish to pose the more specific question: for *which* fields F is $(E, \mathscr{E})$ linearly representable in a vector space over F? A very simple result in this connection (already referred to, and set as an exercise) is that if $(E, \mathscr{E})$ is linearly representable over F (i.e. in a vector space over the field F) and F is a subfield of F', then $(E, \mathscr{E})$ is linearly representable over F'.

Of particular interest are the 'binary' and 'regular' spaces. An independence space is *binary* if it is linearly representable over $GF(2)$. (So, in fact, we have a characterization of binary spaces in Exercise 3.16. The reader may like to use this exercise to show that the Fano geometry is binary.) It is an easy consequence of Theorems 5.6 and 5.7 that duals and minors of binary spaces are themselves binary. Also, as we saw earlier, the cycle and cutset spaces are binary. An example of a non-binary space is the truncation at two of the universal structure on a set of four members, although this space is certainly linearly representable (see the exercises). In fact an independence space is known to be binary if and only if it does not have this particular truncation as a minor.

An independence space is *regular* if it is linearly representable over every field. So, again from Theorems 5.6 and 5.7, duals and minors of regular spaces are themselves regular. Examples of regular spaces are the cycle and cutset spaces (see the exercises), whereas the Fano geometry is not regular. In fact, a binary independence space is regular if and only if it does not contain as a minor the Fano geometry or its dual.

As we have already observed, the transversal spaces are linearly representable. The particular kind of representation which emerged from one of our proofs of Theorem 4.7 was over a field of functions. A stronger result is that the transversal spaces are linearly representable over every sufficiently large field.

In order to pursue a little further the question of linear representability over specified fields, we remind the reader of the notion of the characteristic of a field. Let F be a field, and 0 and 1 its zero and identity elements, respectively. If there exists a positive integer n such that $0 = n \cdot 1 (= 1 + \ldots + 1$ with n summands), then the smallest such positive integer is called the *characteristic* of F; it is known to be a prime number. When no such positive integer exists, F is said to have *characteristic zero*. Now given an independence space $(E, \mathscr{E})$, the *characteristic set* $(\mathscr{C}(\mathscr{E})$ consists of those numbers n for which $(E, \mathscr{E})$ is linearly representable over some field of characteristic n. Since n is either prime or zero, it follows that $\mathscr{C}(\mathscr{E})$ is a subset of $P \cup \{0\}$, where P is the set of primes. We give a sample of known results concerning characteristic sets. A very general (and as yet unsolved) problem is to determine precisely which subsets of $P \cup \{0\}$ can occur as characteristic sets.

(a) If $(E, \mathscr{E})$ is a transversal space or a cycle space or the dual of either, then $\mathscr{C}(\mathscr{E}) = P \cup \{0\}$. (For transversal spaces this can be deduced from the vectorial proof of Theorem 4.7; and for cycle spaces it follows from the fact that they are regular.)

(b) Any singleton subset of P is a characteristic set of some independence space. (For example, if $(E, \mathscr{E})$ is the Fano geometry, then $\mathscr{C}(\mathscr{E}) = \{2\}$.)

(c) If $0 \in \mathscr{C}(\mathscr{E})$, then all sufficiently large primes are in $\mathscr{C}(\mathscr{E})$.

(d) If $0 \notin \mathscr{C}(\mathscr{E})$, then $\mathscr{C}(\mathscr{E})$ is a finite set.

The proofs of these assertions are mainly beyond our scope, since they require more technicalities about fields and field extension than we wish to introduce. However, we conclude this section by proving a simpler result related to (c) above.

Theorem 5.10. *Let $(E, \mathscr{E})$ be an independence space which is linearly representable over the field Q of rational numbers. Then $(E, \mathscr{E})$ is also linearly representable over $GF(p)$ (the field of residues modulo p) for all sufficiently large primes p.*

Proof. We write $E = \{e_1, \ldots, e_n\}_{\neq}$ and let $r (> 0)$ equal the rank of $\mathscr{E}$. Then, by Corollary 5.5 there exists an $r \times n$ matrix $M = (m_{ij})$, with each $m_{ij} \in Q$, such that the mapping φ which takes e_i to the ith column of $M (1 \leqslant i \leqslant n)$ is rank-preserving. Further, we may evidently multiply through the matrix by any non-zero integer without affecting this property and so we may assume, without loss in generality, that all the m_{ij} are integers. Let b denote the product of all the non-zero minors of M. We shall show that $(E, \mathscr{E})$ is linearly representable over $GF(p)$ for any prime number p exceeding $|b|$. Let p be such a prime number and for each m_{ij} let $\bar{m}_{ij}$ denote its least non-negative residue modulo p. Let $\bar{M} = (\bar{m}_{ij})$. Since p does not divide b, it follows that M and $\bar{M}$ have precisely the same non-zero minors. So any set of columns of $\bar{M}$ has the same rank (over $GF(p)$) as the corresponding set of columns of M (over Q); and the mapping $\bar{\varphi}$ which takes e_i to the ith column of $\bar{M}$ ($1 \leqslant i \leqslant n$) is rank-preserving. It follows that $(E, \mathscr{E})$ is linearly representable over $GF(p)$.   □

5.5 Some spaces which are not linearly representable

We started our book by reminding the reader about linear independence in a vector space, and thus we motivated the idea of a general independence space. We conclude it now by demonstrating that the class of independence spaces is genuinely larger than the class of those spaces representable in vector spaces. It is indeed, in principle, quite simple to construct independence spaces which are not linearly representable over any field. For example, the direct sum of independence spaces whose characteristic sets are, respectively, the singleton sets $\{p\}$, $\{p'\}$ (see (b) in the previous section), where p, p' are distinct primes, cannot be linearly representable over any field. Examples of non-representable spaces of rank three on sets of nine and ten elements exist which are closely related to the well-known configurations of Pappus and Desargues, and these will be of interest to readers with some knowledge of the projective geometry of the plane. However, the example which we describe below, the 'Vámos space', is a non-representable space on a set of least cardinality possible, namely eight.

Example

Let $E = \{1, 2, 3, 4, 5, 6, 7, 8\}$ and let $\mathscr{E}$ be of rank four and have for its bases all subsets of E of cardinality four with the exception of $\{1, 2, 3, 4\}$, $\{1, 2, 5, 6\}$, $\{1, 2, 7, 8\}$, $\{3, 4, 5, 6\}$ and $\{3, 4, 7, 8\}$. It is readily seen that $(E, \mathscr{E})$ is indeed an independence space. Suppose that the mapping $\varphi : E \to V$ provides a linear representation of $(E, \mathscr{E})$, where V is a vector space over the field F. In particular, then, we have linear relations of the form

$$\varphi(1) + p_1\varphi(2) + q_1\varphi(3) + r_1\varphi(4) = 0,$$
$$\varphi(1) + p_2\varphi(2) + q_2\varphi(5) + r_2\varphi(6) = 0,$$
$$\varphi(1) + p_3\varphi(2) + q_3\varphi(7) + r_3\varphi(8) = 0,$$

where all the coefficients are non-zero elements of F. The assumption $p_1 \neq p_2$ implies that $\varphi(2)$ is linearly dependent upon $\varphi(3)$, $\varphi(4)$, $\varphi(5)$ and $\varphi(6)$ and hence that the sets $\{\varphi(2), \varphi(3), \varphi(4), \varphi(5), \varphi(6)\}$ and $\{\varphi(3), \varphi(4), \varphi(5), \varphi(6)\}$ have the same linear rank; which is false. Therefore $p_1 = p_2$ and, on similar grounds, $p_1 = p_3$. Thus $p_2 = p_3$, and we conclude that the set $\{\varphi(5), \varphi(6), \varphi(7), \varphi(8)\}$ is linearly dependent; which is the desired contradiction.

The geometrical contradiction inherent in the example can be appreciated in terms of affine representability. Indeed, if $(E, \mathscr{E})$ were linearly representable, then, by our earlier remarks, it would also be affinely representable by means of a mapping $\psi : E \to V$, where V is some vector space of dimension 3. The image of E would then consist

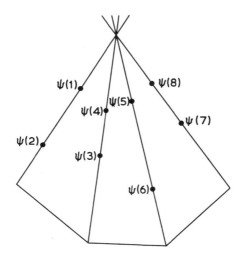

of eight points on the edges of a pyramid, as illustrated (or, possibly, a prism); and this would necessarily imply that the set $\{\psi(5), \psi(6), \psi(7), \psi(8)\}$ was affinely dependent.

Exercises

5.1 Show that neither a minor nor a truncation of a transversal space need itself be a transversal space.

**5.2 Prove that transversal spaces are base-orderable.

5.3 Show that the cycle space of the complete graph on four vertices is not base-orderable (and hence this space is not a transversal space).

5.4 Verify that the truncation at two of the universal structure on a set of three elements is linearly representable over $GF(2)$ but is not affinely representable over $GF(2)$.

5.5 Let $(E, \mathscr{E})$ be an independence space which is linearly representable over a field F, and let F be a subfield of F' (i.e. F' is an extension of F). Show that $(E, \mathscr{E})$ is also linearly representable over F'.

5.6 Show that the truncation at two of the universal structure on a set of four elements is linearly representable but is not binary.

5.7 Let $G = (V, E)$ be a graph with $V = \{v_1, \ldots, v_m\}_{\neq}$ and $E = \{e_1, \ldots, e_n\}_{\neq}$. Let F be a field with zero element denoted by 0, identity element by 1, and the additive inverse of 1 by -1. Define an $m \times n$ matrix $M = (m_{ij})$ by the rule

$$m_{ij} = \begin{cases} 1 & \text{if } e_j \text{ joins } v_i \text{ to } v_k \text{ for some } k > i \\ -1 & \text{if } e_j \text{ joins } v_i \text{ to } v_k \text{ for some } k < i \\ 0 & \text{otherwise.} \end{cases}$$

[This is clearly a generalization of the concept of the incidence matrix of G.] Show that the mapping $\varphi : E \to F^m$ which takes e_i to the ith column of M is rank-preserving with respect to the cycle structure $\mathscr{E}(G)$ on E. Deduce that cycle spaces and cutset spaces are regular.

*5.8 Let E be a set of cardinality n, let m be a positive integer not exceeding n, and let $\mathscr{E}$ be the truncation at m of the universal structure on E. Show that $(E, \mathscr{E})$ is linearly representable over any field F for which $|F| \geq n$.

5.9 Let $(E, \mathscr{E}_1)$, $(E, \mathscr{E}_2)$ be independence spaces with $E_1 \cap E_2 = \varnothing$ and $\mathscr{C}(\mathscr{E}_1) \cap \mathscr{C}(\mathscr{E}_2) = \varnothing$ (where $\mathscr{C}(\mathscr{E}_i)$ denotes the characteristic set of $\mathscr{E}_i$). Show that the independence space $(E_1 \cup E_2, \mathscr{E}_1 \oplus \mathscr{E}_2)$ is not linearly representable over any field.

Hints and solutions to the exercises

Chapter One

1.1 We note first that, if X is a linear subspace of V and $x_1, \ldots, x_n \in X$, then an easy consequence of the definition is that $\alpha_1 x_1 + \ldots + \alpha_n x_n \in X$ for all $\alpha_1, \ldots, \alpha_n \in F$. So, if X is a linear subspace with a basis $\{x_1, \ldots, x_n\}_{\neq}$ and $v \in V \setminus X$, then v is not of the form $\alpha_1 x_1 + \ldots + \alpha_n x_n$; and it is straightforward to check that $\{v, x_1, \ldots, x_n\}_{\neq}$ is linearly independent. Hence, for a linear subspace X, $v \in V \setminus X$ implies that $\dim(X \cup \{v\}) > \dim X$.

Conversely, suppose that $\dim(X \cup \{v\}) > \dim X$ whenever $v \in V \setminus X$. Now, for $x, y \in X$ and $\alpha, \beta \in F$ it is easy to check that $\dim(X \cup \{\alpha x + \beta y\}) = \dim X$ and hence that $\alpha x + \beta y \in X$. So X is a linear subspace of V.

1.2 $\{u_1, \ldots, u_m\}_{\neq}$ is affinely independent
$\Rightarrow$ if $\alpha_1 u_1 + \ldots + \alpha_m u_m = 0$ and $\alpha_1 + \ldots + \alpha_m = 0$, then $\alpha_1 = \ldots = \alpha_m = 0$
$\Rightarrow$ if $\alpha_1(u_1 - u) + \ldots + \alpha_m(u_m - u) = 0$ and $\alpha_1 + \ldots + \alpha_m = 0$, then $\alpha_1 u_1 + \ldots + \alpha_m u_m = \alpha_1 u + \ldots + \alpha_m u = 0$ and $\alpha_1 = \ldots = \alpha_m = 0$
$\Rightarrow$ $\{u_1 - u, \ldots, u_m - u\}$ is affinely independent.

1.3 $\{u_1, \ldots, u_m\}_{\neq}$ is affinely independent
$\Leftrightarrow$ if $\alpha_1 u_1 + \ldots + \alpha_m u_m = 0$ and $\alpha_1 + \ldots + \alpha_m = 0$, then $\alpha_1 = \ldots = \alpha_m = 0$
$\Leftrightarrow$ if $\alpha_1 u_1 + \ldots + \alpha_{m-1} u_{m-1} + (-\alpha_1 - \ldots - \alpha_{m-1})u_m = 0$, then $\alpha_1 = \ldots = \alpha_{m-1} = 0$

117

$\Leftrightarrow$ if $\alpha_1(u_1 - u_m) + \ldots + \alpha_{m-1}(u_{m-1} - u_m) = 0$,   then $\alpha_1 = \ldots = \alpha_{m-1} = 0$

$\Leftrightarrow \{u_1 - u_m, \ldots, u_{m-1} - u_m\}_{\neq}$ is linearly independent.

1.4 If $X = \phi$, then the result is easy. So assume that $X \neq \phi$. We note before proceeding that, for $Y \subseteq V$ with $0 \in Y$,

$$\dim Y = \text{aff rank } Y - 1.$$

So now,

X is an affine subspace of V

$\Leftrightarrow$ "$X - u$" $(= \{x - u : x \in X\})$ is a linear subspace of V (for each $u \in X$)

$\Leftrightarrow \dim((X - u) \cup \{w\}) > \dim(X - u)$ (for each $u \in X$ and $w \in V \backslash (X - u)$)

$\Leftrightarrow$ aff rank $((X - u) \cup \{w\}) >$ aff rank $(X - u)$

$\Leftrightarrow$ aff rank $(X \cup \{w + u\}) >$ aff rank X

$\Leftrightarrow$ aff rank $(X \cup \{v\}) >$ aff rank X (for each $v \in V \backslash X$)

1.5 For the second part, note that if $\dim V = n$, then there is no real loss of generality in taking $V = F^n$.

Chapter Two

2.1 I(1) is immediate. I(2) need only be checked for $A, B \in \mathscr{E}$ with $|A| = 2, |B| = 3$ and $|B \backslash A| \geqslant 2$ (so that $A \nsubseteq B$). But for any A with $|A| = 2, A \cup \{b\} \notin \mathscr{E}$ for at most one b; and I(2) follows.

2.3 We must prove the replacement property I(2). Let, then, $X, Y \in \mathscr{E}$ with $|Y| = |X| + 1$ and take $A = X \cup Y$. Now X is not a maximal member of $\mathscr{E}$ in A (since $Y \subseteq A$, $Y \in \mathscr{E}$ and $|Y| > |X|$). Therefore there exists $y \in A \backslash X (= Y \backslash X)$ such that $X \cup \{y\} \in \mathscr{E}$.

2.4 (i) Let $x|A$ and let D be a maximal independent subset of A. Then either $x \in A$, in which case $x \in D$ or $\{x\} \cup D \notin \mathscr{E}$, and so certainly $x|D$; or $x \notin A$ and $\{x\} \cup B \notin \mathscr{E}$ for some independent subset B of A. Now B is contained in some maximal independent subset D' of $\{x\} \cup A$. Clearly $x \notin D'$ and so D' is a maximal independent subset of A, and $|D'| = |D|$. Since all maximal independent subsets of $\{x\} \cup A$ have the same cardinality, it follows that D is also a maximal independent subset of $\{x\} \cup A$. Thus $\{x\} \cup D \notin \mathscr{E}$ and $x|D$.

(ii) D(1): immediate from the definition of 1.

D(2): if $x = y$, there is nothing to prove. So assume that $x \neq y_1, x|\{y_1, \ldots, y_m\}_{\neq}$ and $x \nmid \{y_2, \ldots, y_m\}$. Then $\{x\} \cup B \notin \mathscr{E}$ for some independent set $B \subseteq \{y_1, \ldots, y_m\}$ but $\{x\} \cup C$ is independent for all independent sets $C \subseteq \{y_2, \ldots, y_m\}$. In particular $y_1 \in B$ and $B' = \{x\} \cup (B \backslash \{y_1\})$ is independent. Hence $\{y_1\} \cup B'$ $(= \{x\} \cup B) \notin \mathscr{E}$ for some independent set $B' \subseteq \{x, y_2, \ldots, y_m\}$, and $y_1 | \{x, y_2, \ldots, y_m\}_{\neq}$ as required.

D(3): we remark first that if $x|A$ and $A \subseteq B$, then evidently $x|B$. Now let $x, y_1, \ldots, y_m, z_1, \ldots, z_n$ be as stated and let Z be a maximal independent subset of $\{z_1, \ldots, z_n\}$. Then, by (i) above, $y_k|Z$ for $1 \leqslant k \leqslant m$ and so clearly $\{w\} \cup Z$ is dependent for each $w \in \{y_1, \ldots, y_m, z_1, \ldots, z_n\} \backslash Z$; i.e. Z is a maximal independent subset of $\{y_1, \ldots, y_m, z_1, \ldots, z_n\}$. Now $x|\{y_1, \ldots, y_m\}$ and so, by our first remark, $x|\{y_1, \ldots, y_m, z_1, \ldots, z_n\}$ and, by (i), $x|Z$. Finally, therefore, $x|\{z_1, \ldots, z_n\}$.

2.5 If $x|A$ and $x \notin A$, then $\{x\} \cup B \notin \mathscr{E}$ for some independent $B \subseteq A$. Hence there exists a circuit $C \subseteq \{x\} \cup B$, and clearly $x \in C \subseteq \{x\} \cup B \subseteq \{x\} \cup A$ as required. Conversely, the existence of such a C implies that $C \backslash \{x\} \in \mathscr{E}$, $C \backslash \{x\} \subseteq A$ and $C \notin \mathscr{E}$, so that $x|A$.

2.6 By similar reasoning to that used in the solution to Exercise 2.1, we need only check I(2) for $A, B \in \mathscr{E}$ with $|A| = 4$, $|B| = 5$ and $|B \backslash A| \geqslant 2$. But I(2) holds for such sets since no set of cardinality four is contained in more than one of the four excluded sets.

2.8 (ii) $\{3\}, \{4\}, \{8\}$ (Each is the complement of a basis of $\mathscr{E}|\{1, 2, 5, 6, 7\}$ in a basis of $\mathscr{E}$).

2.9 Let $\{u_1, \ldots, u_k\}_{\neq}$ be a basis of U. Then $\{[v_1], \ldots, [v_r]\}_{\neq}$ is linearly independent in V/U
$\Leftrightarrow \lambda_1[v_1] + \ldots + \lambda_r[v_r] = [0]$ implies $\lambda_1 = \ldots = \lambda_r = 0$
$\Leftrightarrow \lambda_1 v_1 + \ldots + \lambda_r v_r \in U$ implies $\lambda_1 = \ldots = \lambda_r = 0$
$\Leftrightarrow \lambda_1 v_1 + \ldots + \lambda_r v_r = \mu_1 u_1 + \ldots + \mu_k u_k$ implies
$\lambda_1 = \ldots = \lambda_r = 0$ (and hence $\mu_1 = \ldots = \mu_k = 0$)
$\Leftrightarrow \{v_1, \ldots, v_r, u_1, \ldots, u_k\}_{\neq}$ is linearly independent in V
$\Leftrightarrow \{v_1, \ldots, v_r\} \in \mathscr{E}_{\otimes V \backslash U}$

2.10 (iii) If C_1, C_2 are different circuits in $B \cup \{y\}$, then by C(2)' there is a circuit contained in $(C_1 \cup C_2) \backslash \{y\}$ $(\subseteq B)$. This is clearly impossible.

(iv) If $x \notin C$, then $C \subseteq (B \cup \{y\}) \backslash \{x\}$ and $(B \cup \{y\}) \backslash \{x\} \notin \mathscr{E}$. Conversely, if $(B \cup \{y\}) \backslash \{x\}$ is not a basis of $\mathscr{E}$, then, by a cardinality argument, it is dependent and contains a circuit. By (iii) this circuit must be C, and hence $x \notin C$.

2.11 If $x \in B \cap B'$, then we may choose $y = x$. Let us assume then that $x \in B \backslash B'$, and let C be the unique circuit in $B' \cup \{x\}$. For each

$y \in B' \backslash B$ there exists a circuit $C_y \subseteq B \cup \{y\}$. Now, by Exercise 2.10 (iv), for such a y, $(B' \cup \{x\}) \backslash \{y\}$ is a basis if and only if $y \in C$, and $(B \cup \{y\}) \backslash \{x\}$ is a basis if and only if $x \in C_y$. *So we must show that there exists $y \in C \cap (B' \backslash B)$ such that $x \in C_y$.* Let C^* be a circuit in $B \cup C$ with $x \in C^*$ so chosen that $|C^* \cap (C \backslash B)|$ is minimal. Then there exists $y \in C^* \backslash B$; and we claim that y is as required. For, if $x \notin C_y$, then the existence of a circuit C' with $x \in C' \subseteq (C_y \cup C^*) \backslash \{y\}$ contradicts the minimality of C^*.

2.12 Let D be a maximal independent subset of $A \cap A'$ and let B be a basis of $\mathscr{E}$ with $D \subseteq B \subset A$. Then any $a \in A \backslash B$ will give the required result.

Now let $P, P' \in \mathscr{E}^*$ with $|P'| > |P|$. Then from the above result applied to $A = E \backslash P$ and $A' = E \backslash P'$, we conclude that $P \cup \{a\} \in \mathscr{E}^*$ for some $a \in A \backslash A' = P' \backslash P$.

2.13 It is straightforward to check that the set $B = \{x \in E : \rho(A \cup \{x\}) = \rho(A)\}$ is such that $\rho(B) = \rho(A)$ and is a flat containing A. Also, if F, F' are flats with $F \subset F'$, then $\rho(F) < \rho(F')$. The required results now follow since, if X is the span of A, then $A \subseteq X \subseteq B$ with equality of rank throughout.

2.14 (i) Note first that an intersection of hyperplanes is an intersection of flats, and hence is itself a flat.

Conversely, let $A \subset E$ be a flat with maximal independent subset $\{x_1, \ldots, x_r\}_{\neq}$ $(0 \leqslant r \leqslant n-1)$. Extend this set to $\{x_1, \ldots, x_{n-1}\}_{\neq} \in \mathscr{E}$ whose span, H say, has rank $n-1$ and is thus a hyperplane. Now, if $x \notin H$, then $\{x\} \cup H$ has rank n and so $\{x, x_1, \ldots, x_{n-1}\} \in \mathscr{E}$, $\{x, x_1, \ldots, x_r\} \in \mathscr{E}$ and $x \notin A$. Thus the hyperplane H contains A. Next, let I be the intersection of all hyperplanes containing A. Assume that there exists $y \in I \backslash A$; then $\{y, x_1, \ldots, x_r\} \in \mathscr{E}$, and there exists a basis $\{y, x_1, \ldots, x_r, y_{r+1}, \ldots, y_{n-1}\}$ of $\mathscr{E}$. Exactly as above, the span of $\{x_1, \ldots, x_r, y_{r+1}, \ldots, y_{n-1}\}$ is a hyperplane containing A but not I (since y is not in the span.) This contradiction shows that $A = I$, an intersection of hyperplanes.

(ii) H is a hyperplane of $\mathscr{E}$

$\Leftrightarrow H$ contains no basis of $\mathscr{E}$ and, given any $x \in E \backslash H$, the set $\{x\} \cup H$ contains a basis of $\mathscr{E}$

$\Leftrightarrow E \backslash H$ is contained in no basis of $\mathscr{E}^*$ and, given any $x \in E \backslash H$, the set $E \backslash (\{x\} \cup H)$ is contained in a basis of $\mathscr{E}^*$

$\Leftrightarrow E \backslash H$ is a circuit of $\mathscr{E}^*$.

2.15 Let $(E, \mathscr{E})$ be connected and let $x, y \in E$ with $x \neq y$. Then $\{x, y\} \subseteq C$ for some circuit C, and $C \backslash \{y\} \subseteq B$ for some basis B.

With the help of Exercise 2.10 (iv), it is straightforward to check that $B \setminus \{x\}$ spans a hyperplane H with $x, y \in E \setminus H$. By Exercise 2.14 (ii), $E \setminus H$ is a circuit of $\mathscr{E}^*$; and the result follows.

2.16 Write $E \setminus C^* = H$ (a hyperplane in $\mathscr{E}$). If $C \cap C^* = \{x\}$, then $C \setminus \{x\} \subseteq H$ whereas $C \not\subseteq H$; this contradicts the fact that H is a flat.

2.18 The structure $\mathscr{E}^* | E \setminus E'$ has rank function given by

$$\rho^*(A) = |A| - \rho(E) + \rho(E \setminus A) \quad (A \subseteq E \setminus E')$$

and hence the rank of A in $(\mathscr{E}^* | E \setminus E')^*$ is equal to

$$|A| - \rho^*(E \setminus E') + \rho^*((E \setminus E') \setminus A) \quad (A \subseteq E \setminus E');$$

i.e.

$$|A| - (|E \setminus E'| - \rho(E \setminus E') + \rho(E)) \\ + (|E \setminus E'| - |A| - \rho(E) + \rho(A \cup E')).$$

But this simplifies to $\rho(A \cup E') - \rho(E')$, which is precisely the rank of A in $\mathscr{E}_{\otimes E \setminus E'}$.

2.20 If $(E, \mathscr{E})$ is of rank r and is a proper truncation, then it is a truncation of some space $(E, \mathscr{E}')$ of rank $r + 1$. It is easy to check that bases of $\mathscr{E}'$ are circuits of $\mathscr{E}$, that C, C' are (respectively) the unique bases of $\mathscr{E}'$ containing B, B', and then that, if $e \notin C'$ say, the sets B and C' fail to satisfy I(2) in $\mathscr{E}'$.

2.21 If $(E, \mathscr{E})$ is of rank r and $(E', \mathscr{E}')$ is any independence space of rank at least 1 with $E \cap E' = \phi$, then $\mathscr{E}$ is the restriction to E of the proper truncation at r of $\mathscr{E} \oplus \mathscr{E}'$.

2.22 (i) If $(E, \mathscr{E})$ has rank r and is a truncation of the rank $-(r + 1)$ space $(E, \mathscr{E}')$ then the bases of $\mathscr{E}'$ are circuits of $\mathscr{E}$; and the result follows easily.

(ii) Apply the result of Exercise 2.20 to the bases $B = \{1, 2, 5, 6, 8\}$ and $B' = \{2, 3, 5, 6, 7\}$.

2.23 Let V be a vector space of dimension n and let $\{v_1, \ldots, v_n\}$ be a basis of V. Then, for $n \geqslant 2$, $\{v_1, \ldots, v_n, v_1 + \ldots + v_n\}_{\neq}$ is a circuit of V, and hence V is a circuit space.

Now assume that $n > 2$ and that V is a truncation of an independence space $(V, \mathscr{E})$ of rank $n + 1$. Then $A = \{v_0, v_1, \ldots, v_n\}_{\neq} \in \mathscr{E}$ for some $v_0 = \alpha_1 v_1 + \ldots + \alpha_n v_n$, where $\alpha_1, \ldots, \alpha_n$ are scalars in the underlying field. Since A is a circuit of V it follows that the α_is are all non-zero and that $B = \{v_0, v_0 - \alpha_2 v_2, \ldots, v_0 - \alpha_n v_n\}$ is a basis of V. Now apply I(2) to A, $B \in \mathscr{E}$ to give $C = \{v_0, v_0 - \alpha_2 v_2, \ldots, v_0 - \alpha_n v_n, v_i\}_{\neq} \in \mathscr{E}$ for some i, $1 \leqslant i \leqslant n$. Since $n > 2$ it follows that $\{v_0, v_0 - \alpha_j v_j, v_i\}_{\neq}$

is linearly independent in V for $2 \leqslant j \leqslant n$. Clearly this implies that $i = 1$ and that $C = \{v_0, v_1, v_0 - \alpha_2 v_2, \ldots, v_0 - \alpha_n v_n\}$. Now we can apply B(2) to A and C in $\mathscr{E}$ and deduce similarly that $\{v_0, v_1, \ldots, v_{n-1}, v_0 - \alpha_n v_n\}_{\neq} \in \mathscr{E}$. Hence $\{v_1, \ldots, v_{n-1}, v_0 - \alpha_n v_n\}_{\neq}$ $(= \{v_1, \ldots, v_{n-1}, \alpha_1 v_1 + \ldots + \alpha_{n-1} v_{n-1}\}_{\neq})$ is a basis of V; and this contradiction shows that V is not a proper truncation.

2.25 If A is a basis of $\mathscr{E}_{\otimes E \setminus E'}$, then $A \cup B$ is a basis of $\mathscr{E}$ for some $B \subseteq E'$. Hence $A \cup B \cup \{x\}$ is a circuit of $\mathscr{E}$ for some $x \in E$. It is straightforward to check that B is a basis of $\mathscr{E} \mid E'$ and hence that $x \notin E'$; it follows that $A \cup \{x\}$ is a circuit of $\mathscr{E}_{\otimes E \setminus E'}$ and that $\mathscr{E}_{\otimes E \setminus E'}$ is a circuit space.

2.26 (i) Clear by finiteness.

(ii) We shall establish just $C(2)'$ for $\mathscr{C}$, since $C(1)$ is immediage. So let C, C' be distinct members of $\mathscr{C}$ and let $x \in C \cap C'$. Then either $(C \cup C') \setminus \{x\}$ contains a circuit of $\mathscr{E}$ of cardinality not exceeding r (which is therefore in $\mathscr{C}_0$) or $(C \cup C') \setminus \{x\}$ contains no such circuit. In the latter case either $|(C \cup C') \setminus \{x\}| = r + 1$ (in which case $C, C' \in \mathscr{C}_N$ and, by II, $(C \cup C') \setminus \{x\} \in \mathscr{C}_N$) or $|(C \cup C') \setminus \{x\}| \geqslant r + 2$ (in which case, by III, $(C \cup C') \setminus \{x\}$ contains a member of $\mathscr{C}$). So $\mathscr{C}$ *is* the collection of circuits of an independence structure, $\mathscr{E}'$ say. By III, $\mathscr{E}'$ has rank less than $r + 2$, and so, as $\mathscr{E} \subseteq \mathscr{E}'$, it follows that $\mathscr{E}'$ has rank r or $r + 1$.

(iii) For $|A| \leqslant r$, $A \in \mathscr{E}$ if and only if $A \in \mathscr{E}'$. So the truncation of $\mathscr{E}'$ at r is precisely $\mathscr{E}$.

(iv) The rank of $\mathscr{E}'$ is r if and only if $\mathscr{E} = \mathscr{E}'$. So if $\mathscr{E} \neq \mathscr{E}'$, then $\mathscr{E}'$ has rank $r + 1$ and, by (iii), $\mathscr{E}'$ properly truncates to $\mathscr{E}$. Conversely, assume that $\mathscr{E} = \mathscr{E}'$ but that $\mathscr{E}$ is a proper truncation (of $\mathscr{E}''$, say, of rank $r + 1$). Then every member of $\mathscr{E}'' \setminus \mathscr{E}'$ will have cardinality $r + 1$ and will be a circuit of $\mathscr{E}'$; so we may choose one, C'' say, in $\mathscr{C}_i$ such that $\mathscr{C}_{i-1}$ contains no circuit of cardinality $r + 1$ of $\mathscr{E}'$ which is a member of $\mathscr{E}''$. So $C'' = (C \cup C') \setminus \{x\}$ for some $C, C' \in \mathscr{C}_{i-1}$ with $C \neq C'$ and $x \in C \cap C'$. By the choice of i, C and C' are circuits of $\mathscr{E}''$ and so $(C \cup C') \setminus \{x\}$ contains a circuit of $\mathscr{E}''$; which is a contradiction. So if $\mathscr{E} = \mathscr{E}'$, then $\mathscr{E}$ is not a proper truncation.

2.28 Suppose $B = \{x_1, \ldots, x_r\}_{\neq}$ is a heaviest basis, with $w(x_1) \geqslant \ldots \geqslant w(x_r)$, which cannot be obtained by the algorithm. Assume that $A_i = \{x_1, \ldots, x_i\}$ has been constructed by the algorithm, but that $A_{i+1} = \{x_1, \ldots, x_{i+1}\}$ is inadmissible. Then there exists $x \in E \setminus \{x_1, \ldots, x_i\}$ with $w(x) > w(x_{i+1})$ and such that $A'_{i+1} = \{x_1, \ldots, x_i, x\} \in \mathscr{E}$. But then any basis B' with $A'_{i+1} \subseteq B' \subseteq A'_{i+1} \cup B$ will be heavier than B.

Next we assume that $B' = \{y_1, \ldots, y_r\}_{\neq}$ is as stated and that $w(x_1) = w(y_1), \ldots, w(x_i) = w(y_i), w(x_{i+1}) < w(y_{i+1})$, say. Since $\{x_1, \ldots, x_i\}, \{y_1, \ldots, y_{i+1}\} \in \mathscr{E}$ there is a j with $1 \leqslant j \leqslant i+1$ such that $\{x_1, \ldots, x_i, y_j\}_{\neq} \in \mathscr{E}$. But then $w(y_j) \geqslant w(y_{i+1}) > w(x_{i+1})$; and $A_{i+1} = \{x_1, \ldots, x_{i+1}\}$ cannot have been produced by the algorithm from $A_i = \{x_1, \ldots, x_i\}$.

2.29 The verifications are easy, by a consideration of cases. The induced independence structures are

$$\mathscr{E}_1 = \{A \subseteq E : |A| \leqslant k\}$$

and

$$\mathscr{E}_2 = \begin{cases} \mathscr{P}(E) & \text{if } k \geqslant 1 \\ \{\phi\} & \text{if } k = 0. \end{cases}$$

2.30 (i) $\mathscr{E}$ is a copy of $\mathscr{E}' | \{y_1, \ldots, y_m\}$.

(ii) $$\mathscr{E} = \begin{cases} \{A \subseteq E : |A| \leqslant r\} & \text{if } m > r \\ \mathscr{P}(E) & \text{if } m \leqslant r, \end{cases}$$

where r is the rank of $\mathscr{E}'$.

2.31 The conditions are that $|Y| \leqslant k \dim Y$ for all $Y \subseteq X$. These conditions hold for X_1 and can be checked by a consideration of cases. One such partition is

$$\{(4, 0), (3, 1)\}, \{(2, 2), (1, 3)\}, \{(3, 3), (0, 4)\}$$

The conditions fail for X_2 (e.g. for $Y = \{(1, 1), (2, 2), (3, 3), (4, 4)\}$).

2.32 (ii) If $A \in \mathscr{E}_1 \cdot \mathscr{E}_2$, then $A \cap (B_1^* \cup B_2^*) = \phi$ for some bases B_1^*, B_2^* of $\mathscr{E}_1^*$, $\mathscr{E}_2^*$, respectively. But then $A \cap B_1^* = A \cap B_2^* = \phi$; and $A \in \mathscr{E}_1 \cap \mathscr{E}_2$.

Counter-example: Let $E = \{1, 2, 3\}$, let $\mathscr{E}_1$ have bases $\{1, 2\}$, $\{1, 3\}$ and let $\mathscr{E}_2$ have bases $\{1, 2\}$, $\{2, 3\}$.

2.33 (i) (a) and (b) are immediate and (c) has been verified in Exercise 2.16.

(d) Let $\{x\}, E_1, E_2$ be as stated but with no $C \in \mathscr{C}$ satisfying $x \in C \subseteq \{x\} \cup E_1$. Then x is not in the span of E_1, and so there is a hyperplane H containing E_1 but excluding x. Hence $D = E \setminus H$ is a member of $\mathscr{D}$ with the required properties.

(ii) We verify $C(2)$ for $\mathscr{C}$. Let C, $C' \in \mathscr{C}$ with $y \in C \cap C'$ and $x \in C \backslash C'$. Apply (d) to $\{x\}$, $E_1 = (C \cup C') \backslash \{x, y\}$ and $E_2 = (E \backslash (C \cup C')) \cup \{y\}$ to give either $C'' \in \mathscr{C}$ with $x \in C'' \subseteq \{x\} \cup E_1$ or $D'' \in \mathscr{D}$ with $x \in D'' \subseteq \{x\} \cup E_2$. But in the latter case $C \cap D'' = \{x\}$ (which is impossible) and in the former case $C'' \subseteq (C \cup C') \backslash \{y\}$ as required.

Hence $\mathscr{C}$ is the collection of circuits of some independence space $(E, \mathscr{E})$ and, similarly, $\mathscr{D}$ is the collection of circuits of $(E, \mathscr{E}')$, say; and it only remains to verify that $\mathscr{E}' = \mathscr{E}^*$. We note that, by property (c), if $x \in E$, $B \subseteq E$ and $C \in \mathscr{C}$ satisfy $x \in C \subseteq \{x\} \cup B$, then there exists no $D \in \mathscr{D}$ with $x \in D \subseteq \{x\} \cup (E \backslash B)$. Therefore, if there exists $C \in \mathscr{C}$ with $x \in C \subseteq \{x\} \cup B$, then $E \backslash B \in \mathscr{E}'$. It follows that, if B is a basis of $\mathscr{E}$, then $E \backslash B \subseteq B'$ for some basis B' of $\mathscr{E}'$. Assume that $E \backslash B \subset B'$ and that $x \in B' \backslash (E \backslash B) = B \cap B'$. Then $\{x\}$, $E_1 = B \backslash \{x\}$ and $E_2 = E \backslash B$ have the properties that $\{x\} \cup E_1 \in \mathscr{E}$ and $\{x\} \cup E_2 \in \mathscr{E}'$ and so they contradict property (d). Hence $E \backslash B = B'$; and every basis of $\mathscr{E}$ is the complement of a basis of $\mathscr{E}'$. Similarly, every basis of $\mathscr{E}'$ is the complement of a basis of $\mathscr{E}$; and so $\mathscr{E}' = \mathscr{E}^*$.

Chapter Three

3.1 We prove this result by induction on $|V|$, the case $|V| = 1$ being trivial. So assume that T is a spanning tree of the connected graph $G = (V, E)$, that $|V| > 1$, and that the result is known for graphs with fewer than $|V|$ vertices. Now (V, T) has a vertex v of degree 1 (for otherwise each vertex has degree at least 2 and (V, T) contains a cycle) with, say, $e = vv' \in T$. The graph $(V \backslash \{v\}, T \backslash \{e\})$ is connected and contains no cycles and so it has a spanning tree $T \backslash \{e\}$ to which the induction hypothesis can be applied to give

$$|T \backslash \{e\}| = |V \backslash \{v\}| - 1$$

and so

$$|T| = |V| - 1$$

3.2 Use Exercise 2.10.

3.3 For $A \subseteq E$ let $c(A)$ denote the number of components of (V, A). Then $\rho(A)$ is the cardinality of a maximal independent subset of A, i.e. of the union of $c(A)$ spanning trees of the components $(V_1, A_1), \ldots, (V_{c(A)}, A_{c(A)})$, say, of (V, A). Therefore, by Exercise 3.1,

$$\rho(A) = (|V_1| - 1) + \ldots + (|V_{c(A)}| - 1)$$
$$= (|V_1| + \ldots + |V_{c(A)}|) - c(A) = |V| - c(A).$$

Further, by Theorem 2.10,

$$\rho^*(A) = |A| - \rho(E) + \rho(E \backslash A)$$
$$= |A| - |V| + c(E) + |V| - c(E \backslash A)$$
$$= |A| + 1 - c(E \backslash A).$$

3.4 Some other examples of graphs which are their own geometric duals are the 'wheels' defined in Exercise 3.19.

3.5 Assume that $K_5 = (V, E)$ is planar with geometric dual $G^* = (V^*, E^*)$. Then

$$|E^*| = |E| = 10 \quad \text{and} \quad |V^*| = |E| - |V| + 2 = 7$$

and so $2|E^*| < 3|V^*|$. Hence some vertex v of G^* has degree less than 3. Since G^* is connected, the edge-set through v disconnects G^* and contains a cutset. But $\mathscr{E}(K_5)$ is a copy of $\mathscr{E}^*(G^*)$ and so K_5 contains a cycle of 2 or fewer edges. This contradiction shows that K_5 is not planar.

3.6 Label the same edges of G as 1 to 7 in each of its planar representations and then draw the duals with edges 1^* to 7^* and $1^{*\prime}$ to $7^{*\prime}$ respectively. Then it is straightforward to check that $\{e_1^*, \ldots, e_r^*\}$ is a cycle of G^* if and only if $\{e_1^{*\prime}, \ldots, e_r^{*\prime}\}$ is a cycle of $G^{*\prime}$.

3.7 Let $G = (V, E)$ have n components $G_1 = (V_1, E_1), \ldots, G_n = (V_n, E_n)$. Then

$$A \in \mathscr{E}(G) \Leftrightarrow A \text{ contains no cycle of } G$$
$$\Leftrightarrow A \cap E_1, \ldots, A \cap E_n \text{ each contains no cycle}$$
$$\Leftrightarrow A \cap E_1 \in \mathscr{E}(G_1), \ldots, A \cap E_n \in \mathscr{E}(G_n)$$
$$\Leftrightarrow A \text{ is a union of members of } \mathscr{E}(G_1), \ldots, \mathscr{E}(G_n)$$
$$\Leftrightarrow A \in \mathscr{E}(G_1) \oplus \ldots \oplus \mathscr{E}(G_n).$$

Thus, if ρ denotes the rank function of $\mathscr{E}(G)$ and ρ_i that of $\mathscr{E}(G_i)$, then

$$\rho(E) = \rho_1(E_1) + \ldots + \rho_n(E_n)$$
$$= (|V_1| - 1) + \ldots + (|V_n| - 1) = |V| - n;$$

i.e. the number of vertices of G minus its number of components.

3.8   With the notation of the above solution,

$A \in \mathscr{E}^*(G) \Leftrightarrow$ the number of components of G is equal to the number of components of $(V, E \backslash A)$

$\Leftrightarrow$ the number of components of $G_i = (V_i, E_i)$ is equal to the number of components of $(V_i, E_i \backslash A)$ $(= 1)$ for $1 \leqslant i \leqslant n$

$\Leftrightarrow A \cap E_1 \in \mathscr{E}^*(G_1), \ldots, A \cap E_n \in \mathscr{E}^*(G_n)$, etc.

Also, $\rho^*(E) = \rho_1^*(E_1) + \ldots + \rho_n^*(E_n)$
$= (|E_1| - |V_1| + 1) + \ldots + (|E_n| - |V_n| + 1)$
$= |E| - |V| + n.$

3.9 $(\mathscr{E}(G))^* = (\mathscr{E}(G_1) \oplus \ldots \oplus \mathscr{E}(G_n))^*$ (by Exercise 3.7)
$= (\mathscr{E}(G_1))^* \oplus \ldots \oplus (\mathscr{E}(G_n))^*$
(by an easy extension of Exercise 2.17)
$= \mathscr{E}^*(G_1) \oplus \ldots \oplus \mathscr{E}^*(G_n)$ (by Corollary 3.5)
$= \mathscr{E}^*(G)$ (by Exercise 3.8)

3.10   We simply verify I(2) for $\mathscr{E}$. So let $A, B \in \mathscr{E}$ with $|A| + 1 = |B|$, and let A' consist of those members of A which lie in a component of (V, A) which contains a cycle. Let $B_1 = B \cap (A \backslash A')$ and $B_2 = B \backslash B_1$. Note that, since $B_2 \in \mathscr{E}$, the number of vertices forming endpoints of edges of B_2 is at least $|B_2|$; and therefore number of endpoints of edges of B_2

$$\geqslant |B_2| = |B| - |B_1| > |A| - |B_1|$$
$$\geqslant |A'| = \text{number of endpoints of edges of } A'.$$

Hence there exists $b \in B_2$ not both of whose endpoints are endpoints of members of A'. It is now easy to check that $b \in B \backslash A$ and $A \cup \{b\} \in \mathscr{E}$.

3.11   (i) Let $A \subseteq E \backslash E'$. Then

$A \notin \mathscr{E}(G)_{\otimes E \backslash E'} \Leftrightarrow A \cup B \notin \mathscr{E}(G)$ for some maximal independent subset B of E'

$\Leftrightarrow A \cup B$ contains a cycle C of G for some maximal independent subset B of E'

$\Leftrightarrow$ there exists a cycle C of G in $A \cup E'$ with $C \cap A \neq \phi$

$\Leftrightarrow$ there exists a cycle of G' in A

$\Leftrightarrow A \notin \mathscr{E}(G').$

(ii) $\mathscr{E}^*(G') = (\mathscr{E}(G'))^* = (\mathscr{E}(G)_{\otimes E \backslash E,})^* = \mathscr{E}^*(G)|E \backslash E'$ (by Exercise 2.18).

3.12 Apart from any direct graph-theoretic proofs of this result, we may note that

A is a disjoint union of cycles

$\Leftrightarrow A \leftrightarrow x \in X$ (Theorem 3.9)

$\Leftrightarrow x \cdot y = 0$ for all $y \in Y$ (Theorem 3.12)

$\Leftrightarrow |A \cap C|$ is even for all cutsets C.

3.13 dim Y = maximum number of linearly independent rows of $J(G)$

= maximum number of linearly independent columns of $J(G)$

= rank of $\mathscr{E}(G) = |V| - 1$.

Therefore,

dim $X = n - (|V| - 1)$ (by Theorem 1.3)

$= n - |V| + 1$

= rank of $\mathscr{E}^*(G)$.

Also,

$$X + Y = V_n \Leftrightarrow \dim(X + Y) = n(= \dim X + \dim Y - \dim(X \cap Y))$$
$$\Leftrightarrow \dim(X \cap Y) = 0$$
$$\Leftrightarrow X \cap Y = \{0\}.$$

3.15 (i) Simply check that there are $|E| - |V| + 1$ such cycles and that the set of these cycles corresponds to a linearly independent subset of X (and hence a basis of X).

(ii) Similarly for Y.

3.16 (ii) $\Rightarrow$ (i) Let X be the subspace of V_n spanned by the vectors which correspond to the circuits of $\mathscr{E}$ and let

$$Y = \{y \in V_n : x \cdot y = 0 \quad \text{for all } x \in X\}$$

(in which case, by Corollary 1.4,

$$X = \{x \in V_n : x \cdot y = 0 \quad \text{for all } y \in Y\}).$$

Let M be an $s \times n$ matrix whose rows form a spanning set of Y. Then $(\mu_1, \ldots, \mu_n) \in X$ if and only if $M \begin{pmatrix} \mu_1 \\ \vdots \\ \mu_n \end{pmatrix} = \begin{pmatrix} 0 \\ \vdots \\ 0 \end{pmatrix}$. So for $A = \{e_1, \ldots, e_r\}$, say,

$A \in \mathscr{E} \Leftrightarrow A$ contains no circuit of $\mathscr{E}$

$\Leftrightarrow A$ contains no non-empty set of the form $C_1 \vartriangle \ldots \vartriangle C_p$, where $C_1, \ldots, C_p$ are circuits of $\mathscr{E}$ (by (ii))

$\Leftrightarrow A$ contains no non-empty set corresponding to a vector in X

$\Leftrightarrow (\lambda_1, \ldots, \lambda_r, 0, \ldots, 0) \in X$ implies $\lambda_1 = \ldots = \lambda_r = 0$

$$\Leftrightarrow M \begin{pmatrix} \lambda_1 \\ \vdots \\ \lambda_r \\ 0 \\ \vdots \\ 0 \end{pmatrix} = \begin{pmatrix} 0 \\ \vdots \\ 0 \end{pmatrix} \text{ implies } \lambda_1 = \ldots = \lambda_r = 0$$

⇔ the first r columns of M are distinct and form a linearly independent set.

(i) ⇒ (ii) Assume that (i) holds. Then it is straightforward to check that

(a) if C is a circuit of $\mathscr{E}$, then the sum of the columns of M corresponding to the members of C is equal to the zero vector;

(b) if the sum of the columns of M corresponding to the members of a set C_i is equal to the zero vector for $1 \leqslant i \leqslant p$, then the sum of the columns of M corresponding to the members of $C_1 \vartriangle \ldots \vartriangle C_n$ is also equal to the zero vector.

So if $C_1, \ldots, C_p$ are circuits of $\mathscr{E}$ with $C_1 \vartriangle \ldots \vartriangle C_p \neq \phi$, then the columns of M corresponding to $C_1 \vartriangle \ldots \vartriangle C_p$ add to give zero and so $C_1 \vartriangle \ldots \vartriangle C_p \notin \mathscr{E}$; i.e. $C_1 \vartriangle \ldots \vartriangle C_p$ contains a circuit of $\mathscr{E}$.

3.17 Prove the result first for a graph G without loops or repeated edges; the general result then follows easily. If G is a single cycle, then $(E, \mathscr{E}(G))$ is trivially a circuit space since E is itself a circuit. Conversely, if $(E, \mathscr{E}(G))$ is a circuit space, then, since the edge-set through any vertex of G is a member of $\mathscr{E}(G)$ and hence contained in a circuit, G can have no vertices of degree d, where $d = 1$ or $d \geqslant 3$.

3.18 Note first that Theorem 3.19 (i) still holds for $G = (V, E)$, connected or otherwise. So E can be partitioned into k forests if and only if $k\rho(A) \geqslant |A|$ for all $A \subseteq E$. But if this inequality is only given for connected $A \subseteq E$, then it can easily be extended to any $B \subseteq E$ by a consideration of the sets of edges in the components of (V, B).

3.19 Let A be a non-empty connected set of edges and m the number of its endpoints. Let A' be the set of *all* edges of G with those endpoints. Then $\rho(A) = m - 1$ and if the central vertex of the wheel is not one of the m vertices, then

$$|A| \leqslant |A'| \leqslant m \leqslant 2(m - 1) = 2\rho(A).$$

On the other hand, if the central vertex of the wheel is included in the m vertices, then

$$|A| \leqslant |A'| \leqslant (m-1) + (m-1) = 2\rho(A).$$

3.20 With the same notation as in the solution of Exercise 3.19,

$$\frac{n}{2}\rho(A) = \frac{n}{2}(m-1) \geqslant \frac{m(m-1)}{2} = |A'| \geqslant |A|.$$

3.21 Let A be a connected set of edges and $r' + s'$ the number of its endpoints (with the obvious notation and $1 \leqslant r' \leqslant r$, $1 \leqslant s' \leqslant s$) and let A' be the set of *all* edges of G with these endpoints. Then it is straightforward to check that $rs/(r+s-1) \geqslant r's'/(r'+s'-1)$, and so

$$rs/(r+s-1)\rho(A) \geqslant r's'/(r'+s'-1)\rho(A) = r's' = |A'| \geqslant |A|$$

(with equalities throughout when $A = E$). Therefore

$$\{rs/(r+s-1)\}\rho(A) \geqslant |A| \quad \text{for all } A \subseteq E$$

but

$$\{rs/(r+s-1)\} - 1)\rho(A) < |A| \quad \text{for some } A \subseteq E.$$

Therefore $K_{r,s}$ can be partitioned into $\{rs/(r+s-1)\}$ forests, but, no fewer.

3.22 The conditions are that

$$|A| \leqslant k\rho^*(A) = k(|A| + 1 - c(E \backslash A)) \quad \forall A \subseteq E.$$

For K_5 the inequalities

$$2c(A) + |A| \leqslant 12$$

can easily be established by a consideration of the cases $c(A) = 1, 2, 3, 4$ and 5 in turn. Therefore,

$$2c(E \backslash A) + |E \backslash A| \leqslant 12 \quad \forall A \subseteq E$$

and

$$|A| \leqslant 2(|A| + 1 - c(E \backslash A)) \quad \forall A \subseteq E.$$

So the edges of K_5 can be partitioned into two non-disconnecting sets (but certainly not one).

3.23 The partition $A_1, \ldots, A_k$ of E is as required if and only if each A_i is the complement of a spanning tree of G; i.e. a basis of $\mathscr{E}^*(G)$. So the required conditions are that $\mathscr{E}^*(G)$ has k disjoint bases and that $k\rho^*(E) = |E|$.

Chapter Four

4.1   For the ith boy, let A_i be the set of girls with whom he is acquainted $(1 \leqslant i \leqslant n)$ and apply Hall's theorem.

4.2   Define a family $\mathfrak{A} = (A_1, \ldots, A_n)$ of subsets of $E = \{x_1, \ldots, x_k\}_{\neq}$ by the rule

$$x_j \in A_i \Leftrightarrow m_{ij} = 1,$$

and apply Hall's theorem. [In this situation, M is called an *incidence matrix of* $\mathfrak{A}$.]

4.3   Let $\mathfrak{F} = (F_1, \ldots, F_n)$, where $F_i = \{j \in I : A_i \cap B_j \neq \phi\}$ $(1 \leqslant i \leqslant n)$. Then $\mathfrak{F}$ has a transversal $\{j_1, \ldots, j_n\}_{\neq}$, with each $j_i \in F_i$ say, if and only if $A_i \cap B_{j_i} \neq \phi$ for $1 \leqslant i \leqslant n$; i.e. if and only if $\mathfrak{A}$ and $\mathfrak{B}$ have a common system of representatives. Now apply Hall's theorem to $\mathfrak{F}$.

4.4   Note that, for $I' \subseteq I$,

$$\left| \left\{ j \in I : \left(\bigcup_{i \in I'} A_i \right) \cap B_j \neq \phi \right\} \right| \geqslant |I'|$$

$$\Leftrightarrow \left| \left\{ j \in I : \left(\bigcup_{i \in I'} A_i \right) \cap B_j = \phi \right\} \right| \leqslant n - |I'|$$

$$\Leftrightarrow \left(\bigcup_{i \in I'} A_i \right) \cap \left(\bigcup_{j \in J'} B_j \right) = \phi \text{ implies } |J'| \leqslant n - |I'|$$

$$\Leftrightarrow |I'| + |J'| > n \text{ implies } \left(\bigcup_{i \in I'} A_i \right) \cap \left(\bigcup_{j \in J'} B_j \right) \neq \phi.$$

So the required result follows easily from Exercise 4.3.

4.5   Let D be a set disjoint from E and of cardinality $m - p$. Denote by $\mathscr{E}_1$ the universal structure on D, let $\mathscr{E}_2 = \mathscr{E}_1 \oplus \mathscr{E}$, and let $\mathfrak{A}'$ be the family $(A_1 \cup D, \ldots, A_m \cup D)$. Then $\mathfrak{A}$ has a partial transversal of length p in $\mathscr{E}$ if and only if $\mathfrak{A}'$ has a transversal in $\mathscr{E}_2$. By Theorem 4.6, this happens if and only if

$$\rho_2 \left(\bigcup_{i \in I'} (A_i \cup D) \right) \geqslant |I'| \quad \forall I' \subseteq \{1, \ldots, m\},$$

where ρ_2 is the rank function of $\mathscr{E}_2$. But, for $I' \neq \phi$,

$$\rho_2 \left(\bigcup_{i \in I'} (A_i \cup D) \right) = |D| + \rho \left(\bigcup_{i \in I'} A_i \right);$$

and the required result follows.

4.6 Apply Exercise 4.5 in the case $\mathscr{E} = \mathscr{E}(\mathfrak{B})$, where ρ is given, as in the proof of Theorem 4.16, by

$$\rho(X) = \min_{J' \subseteq \{1, \ldots, n\}} \left\{ \left| \left(\bigcup_{j \in J'} B_j \right) \cap X \right| - |J'| + n \right\}.$$

4.7 Consider the families $\mathfrak{A} = (\{1\}, \{2, 3\})$, $\mathfrak{B} = (\{2\}, \{1, 3\})$ (see also Exercise 2.32).

4.8 Assume first that G has a Hamiltonian path from v_0 to v_n,

$$v_0, e_1, v_1, \ldots, e_n, v_n,$$

say. Then $e_1 \in B_0 \cap A_1, \ldots, e_n \in B_{n-1} \cap A_n$ and $\{e_1, \ldots, e_n\}_{\neq}$ is a common transversal of $\mathfrak{A}$ and $\mathfrak{B}$ in $\mathscr{E}$.

Conversely, assume that $\mathfrak{A}$ and $\mathfrak{B}$ have a common transversal $E' = \{e_1, \ldots, e_n\}_{\neq}$ in $\mathscr{E}$. Then the associated undirected graph (V, E') has the properties that the degrees of v_0 and v_n are 1, that the degrees of all other vertices are 2, and that (V, E') is a tree. It follows that (V, E') consists of an undirected path from v_0 to v_n using all the vertices of V, in the order

$$v_0, v_{i_1}, \ldots, v_{i_{n-1}}, v_n$$

say. Clearly $(v_0, v_{i_1}) \in E'$; but then $(v_{i_2}, v_{i_1}) \notin E'$ and so $(v_{i_1}, v_{i_2}) \in E'$. Continuing in this way, we see that the above ordering of vertices specifies a Hamiltonian path in G.

4.9 If $\mathfrak{A}, \mathfrak{B}, \mathfrak{C}$ have a common transversal, then we may assume, without essential loss in generality, that $(A_1 \cap B_1 \cap C_1, \ldots, A_n \cap B_n \cap C_n)$ has a transversal. It follows that, for $I', J', K' \subseteq \{1, \ldots, n\}$,

$$\left| \left(\bigcup_{i \in I'} A_i \right) \cap \left(\bigcup_{j \in J'} B_j \right) \cap \left(\bigcup_{k \in K'} C_k \right) \right| \geqslant \left| \bigcup_{i \in I' \cap J' \cap K'} (A_i \cap B_i \cap C_i) \right|$$

$$\geqslant |I' \cap J' \cap K'|$$
$$= |I' \cap J'| + |K'| - |(I' \cap J') \cup K'|$$
$$= |I'| + |J'| - |I' \cup J'| + |K'| - |(I' \cap J') \cup K'|$$
$$\geqslant |I'| + |J'| + |K'| - 2n.$$

However, $\mathfrak{A} = (\{1\}, \{2, 3\})$, $\mathfrak{B} = (\{2\}, \{1, 3\})$, $\mathfrak{C} = (\{3\}, \{1, 2\})$ satisfy these conditions without having a common transversal.

[An attempted proof of sufficiency on the lines of the proof of Theorem 4.16 will fail since, as in Exercise 4.7, the common partial transversals of two families of sets do not, in general, form an independence structure.]

4.10 If M is a common transversal of $\mathfrak{A}$ and $\mathfrak{B}$, then $M \in \mathscr{E}$. On the other hand, if M satisfies the given conditions, then in particular for $I' = \phi$ we have

$$\left| \left(\bigcup_{j \in J'} B_j \right) \cap M \right| \geqslant |J'| + |M| - n \quad \forall J' \subseteq \{1, \ldots, n\}$$

and so, by Corollary 4.4, M is a partial transversal of $\mathfrak{B}$ and $M \in \mathscr{E}$. So we may assume throughout that $M \in \mathscr{E}$, in which case the rank function of $\mathscr{E}(M)$ is given by

$$\rho_M(X) = \min_{J' \subseteq \{1, \ldots, n\}} \left\{ \left| \left(\bigcup_{j \in J'} B_j \right) \cap (X \cup M) \right| - |J'| + n \right\} + |X \cap M| - |M|$$

(from Theorem 2.9).

Now note that $\mathfrak{A}$ and $\mathfrak{B}$ have a common transversal T containing M if and only if T is a transversal of $\mathfrak{A}$ which is independent in $\mathscr{E}(M)$. By Theorem 4.6, this happens if and only if

$$\rho_M \left(\bigcup_{i \in I'} A_i \right) \geqslant |I'| \quad \forall I' \subseteq \{1, \ldots, n\}$$

$$\Leftrightarrow \left| \left(\bigcup_{j \in J'} B_j \right) \cap \left(\bigcup_{i \in I'} A_i \cup M \right) \right| - |J'| + n$$

$$+ \left| \left(\bigcup_{i \in I'} A_i \right) \cap M \right| - |M| \geqslant |I'| \quad \forall I', J' \subseteq \{1, \ldots, n\}$$

$$\Leftrightarrow \left| \left(\bigcup_{i \in I'} A_i \right) \cap \left(\bigcup_{j \in J'} B_j \right) \right| + \left| \left(\bigcup_{i \in I'} A_i \cup \bigcup_{j \in J'} B_j \right) \cap M \right|$$

$$\geqslant |I'| + |J'| + |M| - n \quad \forall I' J' \subseteq \{1, \ldots, n\}.$$

4.11 Clearly, if ρ is the rank function of $\mathscr{E}$, then
$\rho(\{x\} \cup (E \backslash A_i)) > \rho(E \backslash A_i)$ for each i and each $x \in A_i$.

4.12 Observe that, if $\mathfrak{A} = (A_1, \ldots, A_{n-1}, A_n)$, where $x \in A_{n-1} = A_n$ and $\mathfrak{A}' = (A_1, \ldots, A_{n-1}, A_n \backslash \{x\})$, then $\mathscr{E}(\mathfrak{A}) = \mathscr{E}(\mathfrak{A}')$.

4.13 Assume that $B = \{b_2, \ldots, b_n\}_{\neq}$ with $b_i \in A_i$ for $2 \leqslant i \leqslant n$ and that $(A_1 \backslash B, A_2, \ldots, A_n)$ is not a presentation of $\mathscr{E}$. Then clearly some $\{a_1, \ldots, a_n\}_{\neq}$ with $a_i \in A_i$ for $1 \leqslant i \leqslant n$ is not a transversal of $(A_1 \backslash B, A_2, \ldots, A_n)$. Evidently $a_1 \in A_1$ and $a_1 \notin A_1 \backslash B$, and so $a_1 \in A_1 \cap B$. We shall show by induction that $\{a_1, \ldots, a_n\} \subseteq B$. So assume that $a_1 = b_2$ (say) and that $a_2 = b_3, \ldots, a_r = b_{r+1}$ (say). We shall deduce that $a_{r+1} \in B$. First, $\{a_{r+1}, a_1, a_2, \ldots, a_r, a_{r+2}, \ldots, a_n\}$ $(= \{a_{r+1}, b_2, \ldots, b_{r+1}, a_{r+2}, \ldots, a_n\})$ is not a transversal of $(A_1 \backslash B, A_2, \ldots, A_n)$, and so $a_{r+1} \notin A_1 \backslash B$. Second, $a_{r+1} \in A_1 \cup B$, for otherwise $B' = \{b_3, \ldots, b_{r+1}, a_{r+1}, b_{r+2}, \ldots, b_n\}_{\neq}$ $(= \{a_2, \ldots, a_{r+1}, b_{r+2}, \ldots, b_n\})$ is a transversal of $(A_2, \ldots, A_n)$ with $b_2 (\in A_1 \cap B)$ replaced by $a_{r+1} (\notin A_1 \cap B)$. But then $|B' \cap A_1| < |B \cap A_1|$; which contradicts the choice of B. Hence $a_{r+1} \notin A_1 \backslash B$, $a_{r+1} \in A_1 \cup B$ and so $a_{r+1} \in B$. It follows by induction that $\{a_1, \ldots, a_n\}_{\neq} \subseteq B$; and this is impossible since $|B| = n - 1$. Hence $(A_1 \backslash B, A_2, \ldots, A_n)$ must, after all, be a presentation of $\mathscr{E}$. The rest of the solution is now straightforward.

4.14 Assume first that A_1 is a circuit of $\mathscr{E}^*$, i.e. that $E \backslash A_1$ is a hyperplane of $\mathscr{E}$. Then, for $x \in A_1$, $(E \backslash A_1) \cup \{x\}$ has rank n and therefore contains a transversal $\{x, x_2, \ldots, x_n\}_{\neq}$ of $\mathfrak{A}$. Clearly, this is not a transversal of $(A_1 \backslash \{x\}, A_2, \ldots, A_n)$. So it follows that, if each A_i is a circuit of $\mathscr{E}^*$, then $\mathfrak{A}$ is a minimal presentation of $\mathscr{E}$.

Conversely, if $\mathfrak{A}$ is a minimal presentation of $\mathscr{E}$, then (by Exercise 4.13) for $1 \leqslant i \leqslant n$ there exists a transversal of $(A_1, \ldots, A_{i-1}, A_{i+1}, \ldots, A_n)$ contained in $E \backslash A_i$. It follows that each $E \backslash A_i$ is a flat of rank at least $n - 1$ (by Exercise 4.11). Since each $A_i \neq \phi$ it follows that each $E \backslash A_i$ is a hyperplane of $\mathscr{E}$, and that each A_i is a circuit of $\mathscr{E}^*$.

4.16 An easy argument by induction on $|A|$ shows that, if each component of (V, A) contains at most one cycle, then there are $|A|$ distinct vertices such that each of them is the endpoint of a different member of A. On the other hand, if some component (V', A') of (V, A) has more than one cycle, then $|V'| < |A'|$, and it is impossible to find $|A|$ vertices of the required type.

Now let $V = \{v_1, \ldots, v_n\}_{\neq}$ and, for $1 \leqslant i \leqslant n$, let $A_i (\subseteq E)$ be the set of those edges with endpoint v_i. Then, by the above, $A \in \mathscr{E}$ if and only if A is a partial transversal of $(A_1, \ldots, A_n)$.

4.18 Let X', Y' be sets with X', Y', V mutually disjoint and

$|X'| = |Y'| = m$. Let the directed graph $G' = (V', E')$ be specified by the equations

$$V' = X' \cup Y' \cup V$$

and

$$E' = E \cup \{(x', x) : x' \in X', x \in X\} \cup \{(y, y') : y \in Y, y' \in Y'\}.$$

Then it is easy to check that every set separating X' from Y' in G contains at least m elements and so, by Theorem 4.20, X' is linked to Y' in G'. The rest follows easily.

4.19 It is straightforward to check that $\mathfrak{A}$ and $\mathfrak{B}$ have a common transversal if and only if $\{1, \ldots, n\}$ is linked to $\{n+1, \ldots, 2n\}$ in G. Also, if $S \subseteq V$ separates $\{n+1, \ldots, 2n\}$ from $\{1, \ldots, n\}$, then evidently

$$\left(\bigcup_{i \in \{1, \ldots, n\} \backslash S} A_i \right) \cap \left(\bigcup_{j \in \{n+1, \ldots, 2n\} \backslash S} B_j \right) \subseteq S \backslash \{1, \ldots, 2n\}.$$

Therefore, if the given conditions are satisfied, then for such an S,

$$|S| = |S \backslash \{1, \ldots, 2n\}| + |S \cap \{1, \ldots, 2n\}|$$

$$\geqslant \left| \left(\bigcup_{i \in \{1, \ldots, n\} \backslash S} A_i \right) \cap \left(\bigcup_{j \in \{n+1, \ldots, 2n\} \backslash S} B_j \right) \right|$$
$$+ |S \cap \{1, \ldots, 2n\}|$$

$$\geqslant |\{1, \ldots, n\} \backslash S| + |\{n+1, \ldots, 2n\} \backslash S| - n$$
$$+ |S \cap \{1, \ldots, 2n\}|$$

$$= n.$$

4.20 Let $\mathscr{E}$ be the given strict gammoid. Note first the symmetry of $\mathscr{E}$, in the sense that $\{1, 2, 3\}$, $\{1, 2', 3'\}$, $\{1, 2'', 3''\}$, $\{3, 3', 3''\}$ are circuits of $\mathscr{E}$ and that all other subsets of cardinality three are bases. Now assume that $\mathscr{E}$ is the undirected strict gammoid of sets linked to subsets of some $B \subseteq V = \{1, 2, 3, 2', 3', 2'', 3''\}$ with $|B| = 3$ in a directed graph $G = (V, E)$ in which $(u, v) \in E$ if and only if $(v, u) \in E$. Since $\{3, 3', 3''\}$ is a circuit of $\mathscr{E}$, it follows from Menger's theorem that for some $\{x, y\}_{\neq} \subseteq V$ all paths from $\{3, 3', 3''\}$ to B in G meet $\{x, y\}$. If $3 \notin \{x, y\}$ and $\{x, y, 3\}_{\neq} \in \mathscr{E}$, then there exists a path from 3 to B avoiding x and y. Hence either $3 \in \{x, y\}_{\neq}$ or $\{x, y, 3\}_{\neq}$ is a circuit of $\mathscr{E}$; similar results apply to $3'$ and $3''$. Since at most one of $\{x, y, 3\}$, $\{x, y, 3'\}$,

$\{x, y, 3''\}$ can be a circuit, it follows that $\{x, y\} \subseteq \{3, 3', 3''\}$, say $\{x, y\} = \{3, 3'\}$. Hence every path from $3''$ to B in G meets 3 or $3'$. It is now straightforward to check that $(3'', v_0) \in E$ for some $v_0 \in V \backslash \{3, 3', 3''\}$ (for otherwise with $(3'', 3) \in E$, say, the independence of $\{1, 2'', 3\}$ would imply the independence of $\{1, 2'', 3''\}$) and that every path from v_0 to B meets 3 or $3'$. So $\{3, 3', v_0\}_{\neq} \notin \mathscr{E}$; which is a contradiction. [Thus, the set of all undirected strict gammoids is *properly* contained in the set of all strict gammoids.]

4.23 (i) Each vertex of G' has degree 1 or 2.

(ii) If $v \in X_1 \backslash X_2$, then, since $X_2 \cup \{v\}$ is not an independent subset of A, it follows that $X_2 \cup \{v\} \not\subseteq V(M_2)$; and so $v \in V(M_1) \backslash V(M_2)$. Conversely, if $v \in V(M_1) \backslash V(M_2)$ and $v \in A$, then, since $X_1 \cup \{v\}$ ($\subseteq V(M_1)$) is an independent subset of A, it follows that $v \in X_1$ and certainly $v \in X_2$. Therefore

$$v \in X_1 \vartriangle X_2 \Leftrightarrow v \in (V(M_1) \vartriangle V(M_2)) \cap A$$
$$\Leftrightarrow \text{the degree of } v \text{ in } G' \text{ is 1, and } v \in A$$
$$\Leftrightarrow v \text{ is an endpoint of a path component of } G',$$
$$\text{and } v \in A.$$

(iii) If $u \in X_2 \backslash X_1$, then u is an endpoint of a path component of G' (as in (ii)). If, in each case, the other end of the path is in $X_1 \backslash X_2$, then $|X_1 \backslash X_2| \geqslant |X_2 \backslash X_1|$; which contradicts the fact that $|X_2| > |X_1|$. Therefore there exists a path component with set of edges P and with endpoints v, w with $v \in X_2 \backslash X_1$ and $w \notin X_1 \backslash X_2$. Note that P consists of a sequence of edges, the first of which (incident with v) is in M_2, the second in M_1, the third in M_2, and so on until the last is in M_2.

So now, if $u \in X_1$ and $u \in V(P)$, then u is an endpoint of an edge from M_2 which must be in $P \backslash M_1$, and so $u \in V(P \vartriangle M_1)$. On the other hand, if $u \in X_1$ and $u \notin V(P)$, then u is an endpoint of a member of $M_1 \backslash P$, and again $u \in V(P \vartriangle M_1)$.

(iv) It is easy to see that $P \vartriangle M_1$ above is a matching in G and so, as $\{v\} \cup X_1 \notin \mathscr{E}$, it follows that $|X_2| \not> |X_1|$. Hence, by symmetry, $|X_1| = |X_2|$ and, by Exercise 2.3, $(V, \mathscr{E})$ is an independence space.

4.24 Let $\mathscr{E} = \mathscr{E}(\mathfrak{A})$, where $\mathfrak{A} = (A_1, \ldots, A_n)$ is a family of subsets of E, and $E \cap \{1, \ldots, n\} = \phi$. As usual, define a bipartite graph (I, Δ, E) by the rule $\Delta = \{i\,e. : e \in A_i\}$. Then $\mathscr{E}$ is the restriction of the matching structure of this graph to the set E.

Chapter Five

5.1 The contraction of $\mathscr{E}(\{1, 2, 7\}, \{3, 4, 7\}, \{5, 6, 7\})$ away from $\{7\}$ is not transversal; nor is the truncation at two of $\mathscr{E}(\{1, 2\}, \{3, 4\}, \{5, 6\})$.

5.2 Let $\mathfrak{A} = (A_1, \ldots, A_n)$ have transversals $A = \{a_1, \ldots, a_n\}$ and $B = \{b_1, \ldots, b_n\}$ such that $a_i, b_i \in A_i$ for $1 \leqslant i \leqslant n$, and define a bijection $\varphi : A \to B$ by the rule $\varphi(a_i) = b_i$ for each i. In particular, if $b_j \in A \cap B$ it follows that $b_j = a_i$ for some i and in that event $\varphi(b_j) = b_i$. Evidently, for any $a \in A$, $\varphi^{r+1}(a)$ will be defined provided that $\varphi^r(a) \in A$. Now let $a \in A \backslash B$, and form the sequence

$$a, \varphi(a), \varphi^2(a), \ldots.$$

The reader will be able to confirm that the injectivity of φ ensures that the terms of this sequence are distinct and, therefore, that it terminates, at $\varphi^m(a)$ say, in $B \backslash A$. Next, define $\theta : A \to B$ by the rule

$$\theta(a) = \begin{cases} a & \text{if } a \in A \cap B \\ \varphi^m(a) & \text{if } a \in A \backslash B \text{ (}m \text{ as above)}. \end{cases}$$

It is straightforward to check that the identical process with the roles of A and B reversed and φ replaced by φ^{-1} will lead to a mapping from B to A which is the inverse of θ. Thus, in particular, θ is a bijection and clearly it only remains to prove that $(A \backslash \{a\}) \cup \{\theta(a)\}$ is a transversal of $\mathfrak{A}$ for each $a \in A \backslash B$. Consider such an a and assume, without essential loss in generality, that

$$a = a_1 \text{ (say)} \in A \backslash B,$$
$$\varphi(a) = b_1 = a_2 \text{ (say)} \in A \cap B,$$
$$\varphi^2(a) = \varphi(a_2) = b_2 = a_3 \text{ (say)} \in A \cup B,$$
$$\vdots$$
$$\varphi^{m-1}(a) = \varphi(a_{m-1}) = b_{m-1} = a_m \text{ (say)} \in A \cap B,$$
$$\varphi^m(a) = \varphi(a_m) = b_m \in B \backslash A.$$

Then
$$(A \backslash \{a\}) \cup \{\theta(a)\} = \{a_2, a_3, \ldots, a_n\} \cup \{b_m\}$$
$$= \{b_1, b_2, \ldots, b_{m-1}, b_m, a_{m+1}, \ldots, a_n\};$$

which is certainly a transversal of $\mathfrak{A}$. [The proof may, of course, be expressed in terms of alternating paths in the usual bipartite graph associated with $\mathfrak{A}$.]

5.3 There is no suitable bijection, for example, between two of the disjoint bases of this space.

5.4 The independence space is a cycle space and so it certainly is binary (i.e. linearly representable over $GF(2)$). Now assume that $\varphi : \{1, 2, 3\} \to (GF(2))^n$ preserves affine rank. Then

$$\alpha\varphi(1) + \beta\varphi(2) + \gamma\varphi(3) = 0$$

for some α, β, $\gamma \in GF(2)$ with $\alpha + \beta + \gamma = 0$. But then precisely two of α, β, γ are equal to 1 and this implies that two of $\varphi(1)$, $\varphi(2)$, $\varphi(3)$ coincide; which is impossible.

5.5 Suppose that the mapping $\varphi : E \to F^n$ gives rise to a linear representation of $(E, \mathscr{E})$. The natural embedding $\psi : F^n \to F'^n$ is evidently rank preserving; and therefore, so also is $\psi \circ \varphi : E \to F'^n$.

5.6 This space is certainly linearly representable, for example by the four vectors $(1, 0)$, $(1, 1)$, $(1, 2)$, $(1, 3)$ of the real plane. However if C, C' are distinct circuits of cardinality three, then $C \triangle C' \in \mathscr{E}$; and so, by Exercise 3.16, the space is not binary.

5.7 We shall show that $\{e_1, \ldots, e_r\}_{\neq}$ contains a cycle if and only if the $\varphi(e_1), \ldots, \varphi(e_r)$ are linearly related, for the rest of the question will then follow easily. So assume firstly that $\{e_1, \ldots, e_r\}_{\neq}$ contains a cycle, $\{e_1, \ldots, e_s\}_{\neq}$ say. Then either $s = 1$ and $\varphi(e_1) = 0$, or $s > 1$ and, say,

$$e_1 = v_{i_1}v_{i_2}, e_2 = v_{i_2}v_{i_3}, \ldots, e_s = v_{i_s}v_{i_{s+1}} (= v_{i_s}v_{i_1}).$$

In this latter case define $\alpha_1, \ldots, \alpha_s$ by the rule

$$\alpha_j = \begin{cases} 1 & \text{if } i_j < i_{j+1} \\ -1 & \text{if } i_j > i_{j+1}. \end{cases}$$

Then it is easy to check that

$$\alpha_1\varphi(e_1) + \ldots + \alpha_s\varphi(e_s) = 0$$

and hence that the $\varphi(e_1), \ldots, \varphi(e_r)$ are linearly related.

Conversely, if $\varphi(e_1), \ldots, \varphi(e_r)$ are linearly related, then we may assume without loss in generality, that

$$\alpha_1\varphi(e_1) + \ldots + \alpha_s\varphi(e_s) = 0$$

for some s and some $\alpha_1, \ldots, \alpha_s \in F$, none of which is 0. Then let

G' be the graph with edge-set $\{e_1, \ldots, e_s\}$ and vertex-set equal to the endpoints of $\{e_1, \ldots, e_s\}$. It is clear that no vertex of G' has degree 1, for if v_i, say, had degree 1 in G' and was an endpoint of e_j, then the ith co-ordinate of

$$\alpha_1 \varphi(e_1) + \ldots + \alpha_r \varphi(e_r)$$

would be α_j. Hence each vertex of G' has degree 2 or more, and $\{e_1, \ldots, e_r\}$ contains a cycle.

5.8 Let F be a field with $|F| \geqslant n$ and let $\theta : E \to F$ be an injective mapping. Further, let $\psi : F \to F^m$ be defined by the rule

$$\psi(\alpha) = (1, \alpha, \alpha^2, \ldots, \alpha^{m-1}) \quad \forall \alpha \in F.$$

Then $\varphi = \psi \circ \theta : E \to F^m$ is evidently injective and, for any distinct $e_1, \ldots, e_m \in E$, the set $\{\varphi(e_1), \ldots, \varphi(e_m)\}_{\neq}$ is linearly independent, since the determinant of the matrix

$$\begin{pmatrix} 1 & \theta(e_1) & \theta(e_1)^2 & \ldots & \theta(e_1)^{m-1} \\ 1 & \theta(e_2) & \theta(e_2)^2 & \ldots & \theta(e_2)^{m-1} \\ \vdots & \vdots & \vdots & & \\ 1 & \theta(e_m) & \theta(e_m)^2 & \ldots & \theta(e_m)^{m-1} \end{pmatrix}$$

is non-zero.

5.9 If $(E_1 \cup E_2, \mathscr{E}_1 \oplus \mathscr{E}_2)$ is linearly representable over the field F, then so are its restrictions $(E_1, \mathscr{E}_1)$ and $(E_2, \mathscr{E}_2)$.

Further reading

H.H. Crapo and G-C. Rota (1970). *On the Foundations of Combinatorial Theory: Combinatorial Geometries.* M.I.T. Press.

F. Harary (1969). *Graph Theory.* Addison–Wesley. Reading, Mass.

L. Mirsky, (1971). *Transversal Theory.* Academic Press. New York and London.

R. von Randow, (1975). *Introduction to the Theory of Matroids, Lecture Notes in Economics and Mathematical Systems* 109. Springer-Verlag. Berlin.

W.T. Tutte (1971). *Introduction to the Theory of Matroids.* Elsevier. New York.

D.J.A. Welsh (1976). *Matroid Theory.* Academic Press. New York and London.

Index